Statistical Thinking in Business
Second Edition

J. A. John
University of Waikato
New Zealand

D. Whitaker
University of Waikato
New Zealand

D. G. Johnson
Loughborough University
United Kingdom

 Chapman & Hall/CRC
Taylor & Francis Group
Boca Raton   London   New York

Published in 2006 by
Chapman & Hall/CRC
Taylor & Francis Group
6000 Broken Sound Parkway NW, Suite 300
Boca Raton, FL 33487-2742

© 2006 by Taylor & Francis Group, LLC
Chapman & Hall/CRC is an imprint of Taylor & Francis Group

No claim to original U.S. Government works
Printed in the United States of America on acid-free paper
10 9 8 7 6 5 4 3 2 1

International Standard Book Number-10: 1-58488-495-9 (Softcover)
International Standard Book Number-13: 978-1-58488-495-8 (Softcover)
Library of Congress Card Number 2005005695

Library of Congress Cataloging-in-Publication Data

John, J. A.
    Statistical thinking in business / J.A. John, D. Whitaker, D.G. Johnson.-- 2nd ed.
      p. cm.
      Rev. ed. of: Statistical thinking for managers. c2001.
    Includes bibliographical references and index.
    ISBN-13: 978-1-58488-495-8 (hardcover : acid-free paper)
    ISBN-10: 1-58488-495-9 (hardcover : acid-free paper)
    1. Mathematical statistics. I. Whitaker, D. (David) II. Johnson, D. G. (David G.), 1943- III. John, J. A. Statistical thinking for managers. IV. Title.

QA276.12.J375 2005
001.4'22--dc22
                                      2005005695

Taylor & Francis Group
is the Academic Division of T&F Informa plc.

Visit the Taylor & Francis Web site at
http://www.taylorandfrancis.com

and the CRC Press Web site at
http://www.crcpress.com

CONTENTS

PREFACE

Many managers, whether in manufacturing or processing industries, in service organisations, government, education, or health, think that statistics is not relevant to their jobs. Perhaps this is because they view statistics as a collection of complicated techniques for which they see little application.

However, all business activities are subject to variability, and hence uncertainty, and managers have to make decisions in this environment. Managers need to understand the nature of variability. Otherwise they end up making wrong or inappropriate decisions that can be very costly to their organisations. As well as understanding variability, managers should understand that good decision making involves the use of meaningful information. Without good information, or data, managers have to resort to gut feelings or hunches, neither of which can be relied on. Statistics tells us how to deal with variability and how to collect and use data so that we can make informed decisions. All managers need to be able to *think statistically* and to appreciate basic *statistical ideas*.

The aim of the book is to introduce you to this way of thinking and to some of these ideas. Statistical thinking should be an integral and important part of any manager's knowledge. H. G. Wells* said that statistical thinking would one day be as necessary for efficient citizenship as the ability to read and write. It should not be regarded as some separate activity done rarely or left to others to do. Most managers would accept that they need a variety of different skills to inform and enhance their decision making. For example, being able to correctly assess the financial implications of a certain decision and being able to see how it might affect staff and colleagues are undeniably important. But these are, in a sense, secondary issues. If you have failed to find the right course of action in the first place by not being able to fully appreciate the significance of the available data, then knowing what this action will cost to implement is of little consequence.

---

* *Mankind in the Making*, Chapman and Hall: London, 1914, p. 204. This is not exactly what Wells wrote, but it is commonly paraphrased in this form.

Unlike many books concerned with statistics for business or managers, this book does not emphasise mathematics or computation. Statisticians use mathematics extensively in their work, and the computer has become an indispensable tool for the analysis of data. However, our purpose is not to set out the mathematical basis of statistics or to give a series of statistical techniques that require considerable computation. Statistical ideas can be understood with the minimum of arithmetical work, and this is what we aim to do in the book. This does not necessarily make the book simple or basic. Recognising how statistics and statistical thinking fit into the overall business picture is important, and it requires careful thought and understanding. For example, being able to calculate a standard deviation or the equation of a line of best fit is not an end in itself. It is just a tool for describing data and is the beginning of a process of understanding and diagnosis, leading to effective decision making.

A thorough statistical analysis can involve complex ideas and extensive computation, for which specialised knowledge and computer programmes are indispensable. At times, therefore, additional assistance from a qualified statistician will be needed, but this need has to be recognised. Knowing what you can confidently do for yourself and where expert advice is needed are key characteristics of effective management.

It will be necessary to draw graphs, construct tables, and do some calculations. However, all these things can be done quite easily on the computer. In the book we make extensive use of the spreadsheet package Microsoft Excel. Spreadsheets are now standard on all computers and are, therefore, readily accessible. They are also widely used in business, and not just for statistical work. If you have not used a spreadsheet package before or are new to Excel, you may find the introduction given on the CD-ROM a useful starting point. In the book you will be shown how to use Excel for analysing and plotting data. It will be well worthwhile working though the examples and exercises involving Excel on a computer. Excel is easy to use; the more familiar you become with it, the easier it will be to work though those sections involving its use.

A few of the graphs and calculations we use in the book cannot be easily done with Excel 2003. In particular, there is no easy way to obtain a stem and leaf plot or a box plot. We have, therefore, provided an add-in that allows you to construct these plots and has a couple of other features. A copy of this *STiBstat* add-in is on the CD-ROM. We suggest you load it into Excel following the instructions given in the introduction to Excel on the CD-ROM.

One of the innovative features of this book is the inclusion of a number of hands-on exercises and experiments. It is another way we hope to encourage a process of learning by involvement. The questions contained within each chapter are there to actively involve you in the learning process by making you think about important issues to check your understanding. The experiments are a part of this same process, providing (we hope!) an active demonstration of a key concept through participation in a meaningful simulation. The most important of these experiments is the *red beads experiment*, which is introduced in Chapter 1 and forms the basis of many of the important ideas introduced in the book. To get the most out of this experiment, it should be run in the classroom so that you can see actual results occurring. If this is not possible, we have provided a spreadsheet version of the experiment that you can run in Excel. A copy of this spreadsheet is on the CD-ROM. Instructions for using the spreadsheet are provided on the opening sheet. If you cannot participate in a classroom experiment, then substitute your own results from the beads spreadsheet.

In later chapters, you will find four other experiments that can be run either in the classroom or, in most cases, on your own. Two of these, the *dice experiment* and the *quincunx experiment*, have spreadsheet alternatives you can get from the CD-ROM and run so that you can simulate the actual experiment for yourself. Again, you can use your own results instead of those produced in the classroom or contained in the book. Each experiment demonstrates a key learning point. Think carefully about what these learning points are, and try to relate them to real situations that you have experienced.

In addition to the experiments, we provide a number of demonstrations of important statistical techniques we find useful in our teaching. These are explained in the appropriate places of the text. They also are available on the CD-ROM.

In the text you will find numerous questions to answer. They are there to encourage you to think about the various issues involved. You will often get a greater understanding by first thinking things through for yourself. We strongly advise that you attempt to answer these questions as you work through each chapter. Suggested answers will be found on the CD-ROM. At the end of each chapter are exercises for you to do. You will also find over 300 multiple choice questions on the CD-ROM, designed to test your understanding of the statistical issues involved. Some of them are relatively simple, but many will require some thought. A few will require simple calculations. You can check your answers either as you do each question or after you have done a number of them.

Most of the examples in the book are based on our experiences working with a range of organisations throughout New Zealand and the United Kingdom or on those of our colleagues and associates. For confidentiality reasons we do not name those organisations, but we have tried to preserve the realism of the examples. Although the context often relates to New Zealand, most, if not all, examples can be put into the context of any other country. We encourage you to think about similar problems or examples in your country.

The material in the book can be used in a number of different ways. All of it can be adequately covered in a one-semester course of about 36–40 hours. Additional tutorial time should also be allowed in which a selection from the exercises at the end of each chapter can be used. We also suggest you attempt some of the multiple choice questions on the CD-ROM. We see no need for separate computer laboratory sessions, for our focus throughout is on developing your understanding of the role statistics plays in business, rather than on the arithmetic detail. The book can be used for a shorter course by skipping through some of the material previously covered elsewhere or by omitting certain topics. We have used it on a short course by assuming prior knowledge of most of the material in Chapter 3 and by omitting Chapters 2, 9, and 13. Alternatively, or in addition, Chapter 10 could be omitted. For example, omitting Chapters 2, 9, 10, and 13 leaves sufficient material for a one-semester course of about 22–25 hours, supplemented by five to ten tutorials or problem classes. This is not to say that these chapters are unimportant but simply that time constraints sometimes mean not everything can be covered. It is better to work carefully through parts of the book than to attempt to cover it all superficially.

Many people have been involved, either directly or indirectly, in the development of this book. We particularly wish to acknowledge the contribution of Tim Ball, a consultant statistician in Wellington, New Zealand. Tim has freely made material available to us and has been a consistent advocate of the importance of statistical thinking in business. We also wish to acknowledge Jocelyn Dale, Maurice Fletcher, Jock MacKay, Bruce Miller, Peter Mullins, and Peter Scholtes, all of whom have made contributions in various ways. We would also like to thank Kirsty Stroud, formerly our editor at CRC Press, for her encouragement and suggestions on this book. Over the years many undergraduate and MBA students as well as managers and workers in industry have been involved in courses based on all or part of this book. Their feedback and suggestions have been constructive and most welcome. Of course, any mistakes and errors are our responsibility alone.

Chapter 1

Variation

1.1 Variation

We live in a variable world. Managers have to make decisions in an environment subject to variability. Without understanding the nature of variability, managers face the danger of making wrong decisions and mistakes that can be very costly to their organisations. Variation exists in all processes, and reducing variation is the key to improving productivity, quality, and profitability. However, as W. Edwards Deming said, "The central problem in management and in leadership ... is failure to understand the information in variation." Statistics tells us how to deal with variation, and how to collect and use data so that we can make informed decisions.

In this chapter we discuss four problems involving variability. In each case, we would like you to think carefully about what you would do if you were the manager involved and to compare what you think with what the manager actually did. Then we shall run Deming's red beads experiment and revisit these examples in light of this experiment. The lessons learned should give you insight into the nature of variability and an indication of what should be done.

1.2 Airport Immigration

The data in Figure 1.1 represent the number of passengers processed by immigration officers at an international airport over a typical 1-hour period when the officers would be expected to be quite busy. The passengers processed were those requiring a visa to enter the country. Management expected their officers to process ten passengers during this period. The manager of immigration services, in reviewing these figures, was concerned about the performance of Colin and was thinking about how best to reward Frank. He decided to send a letter to Colin giving him an official warning about his poor performance.

Immigration officer	Number of passengers processed
Alan	9
Barbara	10
Colin	4
Dave	8
Enid	6
Frank	14

Figure 1.1   Customers processed.

He made it clear that unless his results improved, they would have to consider terminating his employment. He then rang Frank and congratulated him on his good work and said that the company was looking at him as future management material.

Q1.   Did the manager take appropriate action?
Q2.   What would you do, and why?
Q3.   Would you recommend anything different if the data in Figure 1.1 gave the number of new cars sold in one week by five salespeople working for Hyden Car Dealership?

1.3   Debt Recovery

In Albion Cleaning, an office cleaning company in New Zealand, the level of unrecovered debt has been a cause of concern for some time. The data in Figure 1.2 give the percentage of invoices that have been paid within the due month. The company is aiming for a target recovery level of 80%. When the amount of recovered debt is much lower than this (percentages in the low to mid-70s, for instance), the general manager visits all the district offices in New Zealand to remind the district managers of the importance of customers paying on time.

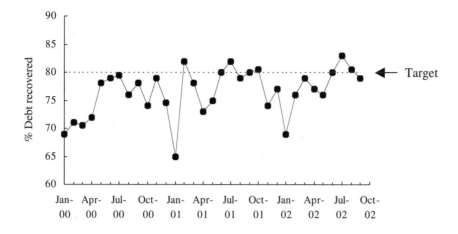

Figure 1.2   Percentage of debts collected within due month.

He tells them their future performance bonuses are at risk unless they quickly reduce the amount of unrecovered debt.

Q4.   Is the general manager taking appropriate action?
Q5.   Has the overall level of debt improved as a result?
Q6.   What would you do, and why?

1.4   Timber Mill Accidents

Joe Birch, the Managing Director of Greenfields Forestry and Forest Products Ltd., was concerned about the number of accidents occurring in the company's timber mills. In an attempt to reduce the number of accidents, Joe wrote a letter to all timber mill workers informing them that anyone involved in an accident in the future would immediately be dismissed from his or her job.

Q7.   Has Joe taken appropriate action?
Q8.   What would you do, and why?

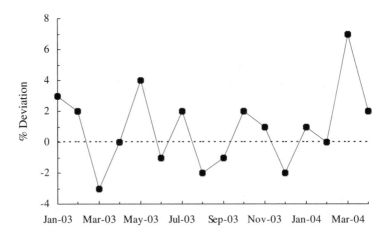

Figure 1.3   Budget deviations.

1.5   Budget Deviations

Jodi Maison, the financial director of Henford Construction Ltd., has been looking at the financial performance of the company over the last year or so. She has collected data on budget deviations, which measure the difference between the amounts budgeted and the actual amounts, expressed as a percentage of the budgeted amounts. The graph in Figure 1.3 gives the budget deviations for a particular account over a 16-month period. The aim, of course, is to have a zero deviation, but some variation is inevitable. Most of the variation lies between –3% and 4%. Jodi was concerned about the 7% deviation in March 2004, so she asked John Allsop, the account manager, to prepare a report explaining why the deviation had been so large that month.

> **Q9.** *How should the financial director have reacted to the 7% deviation in March 2004?*

1.6   Red Beads Experiment

W. Edwards Deming was a prominent guru of quality management. In 1950 he started to teach the Japanese how to improve quality through the use of statistical quality control methods. His management philosophy had a big impact on industry in the United States and elsewhere from 1980 on. Mary Walton, in her book *The Deming Management Method* (Putnam Publishing, 1986), gives a fascinating account of Deming's methods, including his 14 points of management, and devotes a whole chapter to an important experiment on variability that Deming used to carry out in his teaching: the red beads experiment.

We strongly recommend that you experience the beads experiment in a group situation. We run it in a large class using the equipment described below. A number of people get directly involved in the exercise; others take on a managerial role and tell the workers how they can do better! Apart from being very instructive, it is also a fun exercise. If you do not have the equipment, then an alternative is to use the Excel spreadsheet (*Beads.xls*) provided on the CD-ROM. This will allow you to simulate the results from a typical beads experiment.

In the remainder of this section we first describe how to run the experiment using the equipment. Then we explain how to use the beads spreadsheet. The experiment is then run and the data collected, discussed, and plotted. A discussion of the results follows. Finally, we consider what would happen if the experiment were run a large number of times.

✍ *Running the Experiment*

The tools for the experiment are:

- A large number of small plastic beads, the majority of which are white and the others red
- A paddle with 50 bead-size holes in five rows of 10 each, as shown in Figure 1.4
- Two plastic boxes, at least one of which is large enough for dipping the paddle

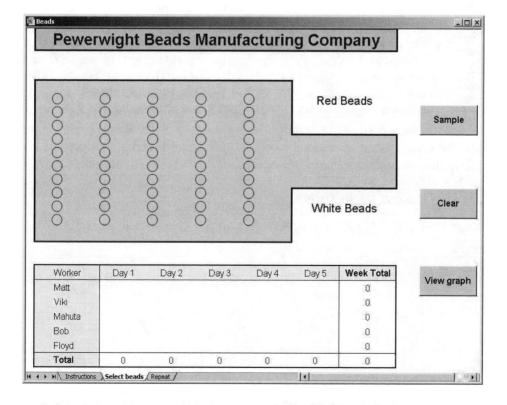

Figure 1.4   Beads experiment spreadsheet.

It is important not to depart from the following procedure for carrying out the experiment.

The beads are first placed in the larger plastic box. They now have to be thoroughly mixed by carefully tipping them into the other box and then tipping them back into the first box. In doing this, each box should be held at the same constant height, say 8 cm, above the other box. A worker then takes the paddle and dips it into the box of beads, being careful to move the paddle along the bottom of the box so that the beads cover all the holes in the paddle. The paddle is removed slowly from the box so that no beads spill out of the paddle. There will now be 50 beads in the paddle, some of which will be white and some red. The number of red beads is counted and recorded. The red beads are defects. The purpose is to produce white beads, not red ones.

Q10. Why is it important to follow the same procedure each time?

✎ Using the Beads Spreadsheet

The Excel spreadsheet *Beads.xls* on the CD-ROM consists of three worksheets. The first sheet gives instructions on how to run the experiment. The experiment is run using the second sheet, which is shown in Figure 1.4.

To dip the paddle into the box of beads and select a sample of 50, click on the *Sample* button to the right of the screen. This will fill the paddle with a mixture of red and white beads. A count of the number of red and white beads appears above and below the paddle handle, and the number of red beads is entered into the first cell of the table. Repeatedly clicking the *Sample* button completes the table of results, day by day. Click on *View graph* to see a run chart of the results. Click the *Clear* button to empty the paddle and clear the table and graph.

The third worksheet allows you to repeat the whole experiment in one go.

✎ Discussion of Results

As the experiment is run, enter the results in the check sheet in Figure 1.5.

Worker	Day 1	Day 2	Day 3	Day 4	Day 5	Week Total
Total						

Figure 1.5   Pewerwight Bead Manufacturing Company.

If you are not experiencing the experiment for yourself, run the beads spreadsheet to get a feel for what the results of the experiment will look like. At the same time, think carefully about what is involved in producing the beads in the paddle. Go through the process in your mind, and try to imagine what is likely to be happening. Finally, enter the data you generate into the table in Figure 1.5.

Now look at the data you have entered in Figure 1.5.

Q11. *What do you notice about the results of the experiment?*

Q12. *What are the smallest and largest numbers of defects produced during the week? Explain why there is so much variability in the results.*

In an attempt to achieve better results (a smaller number of red beads in this case), managers and supervisors often introduce incentive schemes or penalties to get their workers to work more diligently. Slogans, such as "Get it right the first time" are commonly used. Consider whether such methods are likely to be effective in the beads experiment.

Q13. *If an incentive was offered to the best worker on any day, do you think it would help to motivate the workers? Would it be fair to give one of the workers a bonus?*

Q14. *If the worst worker on one day was fired during the experiment, do you think this would be fair? Why, or why not?*

Q15. *In this experiment do you think workers react better to positive or to negative incentives?*

When you experience the experiment, it is very obvious what is really going on. Consider the following questions.

> **Q16.** *What is the fundamental cause of defects in this experiment?*
> **Q17.** *Whose responsibility is it to address this fundamental cause? Is it the workers, or the supervisor, or the inspectors, or management, or who?*
> **Q18.** *What would you do to reduce the number of defects in this experiment?*

✎ *Plotting the Data*

Trends, patterns, changes, and variability in the data are more easily seen in a plot of the data than in a table. Plot the beads data in Figure 1.5 using the chart given in Figure 1.6. Calculate the average number of red beads, and draw a horizontal line on the graph to represent this average. If you used the beads spreadsheet, this line will appear on the graph on the *Select beads* sheet.

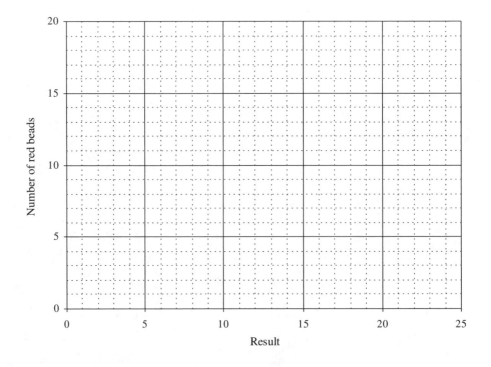

Figure 1.6   Plot of beads data.

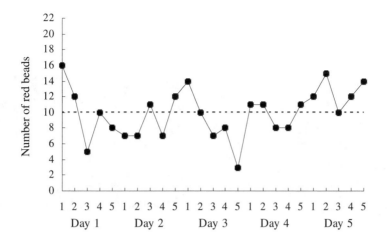

Figure 1.7   Results of first beads experiment.

Q19. Suppose we ran the experiment again. What results would you expect to get? What would the plot of the results look like?

Figures 1.7 and 1.8 show the results of two other experiments carried out with different groups of students, using 1600 beads, of which 1200 are white

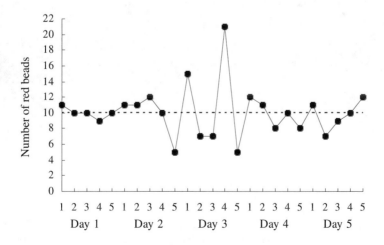

Figure 1.8   Results of second beads experiment.

and 400 are red. The overall average numbers of red beads are also drawn on the plots.

Q20. *What can you say about the results of these two experiments?*
Q21. *In Figure 1.8, what do you think happened at the 4th result on day 3 when 21 red beads were produced?*

Some results will be surprising, while others are not. Deciding which is which is crucial to understanding variation. We discuss this in detail in Chapters 11, 12, and 13.

Q22. *Having now seen the results of the experiment, would you be surprised if someone got*
  a. 50 red beads in the paddle?
  b. 40 red beads?
  c. 30 red beads?
  d. 0 red beads?
Q23. *If we ran the experiment a very large number of times with 1200 white and 400 red beads, what do you think the average number of red beads would be, and why?*
Q24. *Does the evidence given in your plot in Figure 1.6 and in Figures 1.7 and 1.8 support your answer?*

✤ Beads History

A total of 425 results obtained from running the experiment 17 times are plotted in Figure 1.9, using 1600 beads, of which 1200 are white and 400 are red. The average values from each of these experiments are plotted in Figure 1.10. You can also use the third worksheet, called *Repeat* in the *Beads.xls* spreadsheet,

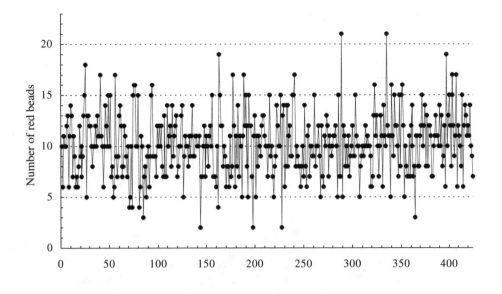

Figure 1.9 Results from 17 beads experiments.

to generate the results of a series of beads experiments and create a graph similar to that in Figure 1.10.

Q25. Is your answer to Q23 consistent with the results given in Figures 1.9 and 1.10? If not, why not?

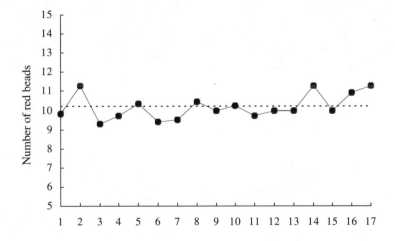

Figure 1.10 Averages from 17 beads experiments.

1.7 Some Conclusions

The beads experiment is, in many senses, a trivial exercise. It is obvious what is going on, especially if you experienced it yourself. In practice, business activities are much more complicated, yet many useful lessons come from this experiment that can be applied to other, more complex, situations. Consider, for instance, the four examples at the beginning of this chapter. None of them is as straightforward or as simple as the beads experiment, yet they have a number of features that closely parallel those in our experiment.

Look again at the airport immigration data in Figure 1.1. Compare these data with the results from one day's production in our experiment.

> **Q26.** *Are they very different? Revisit your answers to the questions in Section 1.2 in light of the beads experiment. What conclusions do you now draw?*
>
> **Q27.** *Suppose four more sets of data are available on the number of passengers processed by the six immigration officers. If Colin still processes the least number of passengers over these five time periods, what conclusions would you draw? What might you do next?*

In the debt recovery example in Section 1.3, the general manager took action when the results were much lower than expected, that is, when the percentage debt recovered was below about 75%.

> **Q28.** *Was this sort of reaction apparent in the beads experiment? Can you see similarities between the two situations? Again review your answers to the questions in Section 1.3.*

In the timber mill accidents example in Section 1.4, Joe was threatening to dismiss any worker involved in an accident.

> **Q29.** *What can we say about the occurrence of accidents in timber mills? What should Joe, as managing director, do before anything else?*

Finally, in the budget deviations example in Section 1.5, the 7% deviation in March 2004 might seem to be a surprising result, but is it?

> *Q30. Were the 21 red beads shown in Figure 1.8 surprising or unusual? How do we know? What do you think?*

1.8 Chapter Summary

In this chapter we have seen how variation has an impact on decision making. Specifically:

- All business activities are subject to variation.
- The red beads experiment showed a process not under the control of the workers, which led to rather large differences in workers' performances. The problem was with the process, not with the people working on the process.
- There are many situations in business that are just like the red beads experiment. Many of the decisions made by managers are largely influenced by the variation in the data and show that they do not understand variation.
- Informed decisions should be based on collecting data and understanding the nature of variation. We look at how to address this in the remainder of the book.

1.9 Exercises

1. A small company launched a new product that turned out to have amazing success. Though there were a few bad months, sales kept increasing, month after month after month. The revenues this product brought in provided much needed relief to the company's cash flow and supported new expansion efforts. Then, suddenly, came a downturn: three bad months in a row. How would you react to the dip in sales in this key product? Would you
   a. Try to motivate your sales staff?
   b. Reduce prices?
   c. Place additional advertisements?
   d. Mail promotional materials to potential new buyers?
   e. Call previous buyers or prospects and try to discover why sales had dropped?
   f. Do nothing?

   (*Source*: From Joiner, B., *Fourth Generation Management,* McGraw-Hill, 1991. With permission.)

2. A psychologist attached to the Israeli Air Force had heard reports that the flight instructors were being abusive to the student pilots. Instead of compliments, there were public reprimands, screaming, and abusive language. On further examination, he found out that the performance of the student

pilots generally improved after they were reprimanded and, indeed, got worse after they were complimented. The flight instructors' behaviour seemed to be vindicated. Based on student pilot performance, he could no longer advocate using more "positive reinforcement" or urge the flight instructors to avoid the harsher responses. The psychologist gave up this project and moved on to other work. However, the results of this study troubled him. During the following year he learned something that led him to reexamine the problem of abusive flight instructors. When he interpreted the results with a different perspective, he found that the flight instructor reprimand did not result in improved performance, nor did compliments result in a worse performance. What was the psychologist's newly acquired perspective?

(*Source*: From Scholtes, P.R., *The Leader's Handbook,* McGraw-Hill, 1998. With permission.)

3. In a fast-food chain, management has introduced a "worker of the month" incentive scheme. Do you think this will motivate staff? Explain the reasons for your answer.

4. Jim, the customer services manager of a large courier company, was concerned about the number of mistakes in customer invoices. He collected some data on the type and frequency of mistakes and found that the biggest problems related to the work of Mary, one of the data input clerks. Jim also discovered that Mary had been warned before about the number of mistakes she was making. Jim is now considering whether to terminate Mary's employment with the company. What other information should Jim consider in making his decision?

5. In the red beads experiment, management has decided to introduce an incentive bonus for any worker who "produces" fewer than five red beads on any day. Do you think this is a good idea? Explain the reason for your answer.

6. Why is the red beads experiment so important in helping managers to understand variation? Why do you think variation is a major problem for most businesses?

7. Rosemary Woods is in charge of customer relations for a large supermarket. Recently she has become concerned about the efficiency of checkout procedures and has collected data on the throughput of all the store's checkout operators over the last month. The data show that 50% of them are working below average, and so she decides to have a word with the "below average" operators. During these discussions she points out that they need to improve their performance to "above average" if they want to continue working for the company. Discuss the merits of Rosemary's actions.

Chapter 2

Problem Solving

2.1 A Scientific Approach to Problem Solving

Knowledge about variation is vitally important, not only for effective decision making, but also for solving problems. Our ability to solve problems quickly and in a sustainable way is directly linked to our ability to understand and interpret variation; otherwise any decisions made may well make things worse or have no effect at all, rather than make things better. In particular, we need to use a *scientific approach,* whereby we learn to:

- Understand variation
- Base decisions on data rather than hunches
- Develop process thinking
- Manage the organisation as a system

The scientific approach involves seeking permanent solutions by looking for root causes of problems, rather than quick fixes that tend to be a reaction to superficial symptoms. This in turn leads to sustainable improvements in *all* areas of business, such as cost control, meeting delivery schedules, employee safety, skill development, supplier relations, new product development, and productivity. It is generally more effective if a number of people work together as a team to pool their knowledge and experience in order to solve problems. Putting in place solutions, through a process of identifying and eliminating root causes, is a more effective approach to problem solving. This is depicted in Figure 2.1.

Unfortunately, problem solving in most organisations tends to be concentrated on "firefighting" obvious and urgent problems, which often takes up a great deal of management time. Even worse, the problems "solved" this way often disappear only for a short time, reappearing in the same or a similar guise a few days, weeks, or months later, to be solved all over again. Many problems are competing for attention, but with firefighting, only enough effort is made

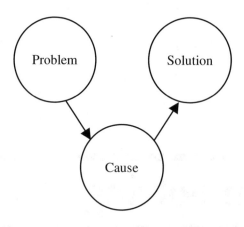

Figure 2.1   The problem–cause–solution approach.

to put out the immediate fires. The problems keep coming back, time and time again. It is rework. This "ready–fire–aim" approach is wasteful and ineffective.

Firefighting omits the cause step. It is an instant solution approach that rarely works. The debt recovery case in the previous chapter is an example. The general manager thought he was being effective, but because he did not understand variability and was not addressing a cause of the problem, the overall level of debt did not improve.

In this chapter we study the scientific approach and use the debt recovery example to illustrate how this approach can effectively solve problems. First, however, we look at why we need data, what we should measure, and some tools that will help us in our problem solving efforts.

2.2   The Need for Data

We have already stated that data are needed for decision making and for understanding and interpreting variation. Data across all business operations are needed for informed decision making, such as data collected from sales records, inventory levels, accounts receivable, daily production records, and defect reports. In many organisations such collection is routine, although the purpose is often not clear and much of it is never used. It is wasteful to obtain more data than is needed to make the appropriate decisions.

For problem solving, some of the reasons for collecting data from a process are:

- To *understand the process*. The system may not be well enough understood for anyone to be able to pinpoint the fundamental causes of the problem. Some basic data are needed in order to know what is going on in the process.
- To *determine priorities*. The problem may have a number of causes. It may therefore be necessary to make decisions about which causes should be examined immediately and which left to some future investigation. This can

be achieved by collecting data on the process so that the potentially important causes can be identified.

■ To *eliminate causes of variation*. For instance, can the overall level of defects be reduced or the variability in the results decreased? We shall see later that to do this it is important to distinguish between two types of causes: special causes and common causes.

■ To *monitor the process*. It may be known what behaviour the process exhibited in the past. Does it still exhibit the same behaviour? Did the policy change instituted two months ago improve the process?

■ To *establish relationships* between different factors or characteristics affecting the process. It may be, for instance, that the variable of interest is difficult or expensive to measure. However, a strong relationship may exist between this variable and another variable that is easier or cheaper to measure. It may then be possible to use this substitute in subsequent analyses.

2.3 Deciding What to Measure

Deciding what to measure is often a very difficult problem. To collect meaningful data we must know precisely what to observe and how to measure it. Consider the following example from Joiner's *Fourth Generation Management* (McGraw-Hill, 1994).

One American airline had a very customer focused goal of making sure flights had on-time departures. Its employees had this goal as one of their most important objectives. The airline put out advertisements to the effect that it was the best airline in terms of on-time departures. Its safety record, baggage handling, and other services were no worse than those of its competitors.

Q1. Suppose you are a regular business traveler, flying most weeks to different cities within the United States. Flight delays with one airline are a great irritation, causing you to be late for meetings or late getting home. Would you be likely to fly with this airline, and why?

Q2. Brian Joiner used to fly regularly with this airline and had exalted frequent flyer status. He flew with them so often that he automatically got a free upgrade to first class on every flight. But their on-time-departure goal upset him so much that he flies with this airline now only when absolutely necessary. What do you think could have upset him?

We also have to be very clear as to the purpose of collecting the data. Problems can occur when the measurements are inconsistent from day to day or from person to person or from machine to machine. Unless we get consistency, difficulties will arise between, for instance, customers and suppliers or people within the same organisation. For example, a salesperson is told that her performance will be judged on the percentage change in this year's sales over last year's sales. What does this mean? Specifically:

- Average percentage change each month? each week? each day?
- For each product?
- Based on sales volume or sales revenue? gross sales? net sales?
- Using last year's prices or this year's prices?

Often customers specify what they want in vague terms that are difficult to define and measure. For example, what is meant by good tasting ice cream, good or bad service in a restaurant, on-time delivery, smoothness, fresh, user-friendly, or easy to assemble?

Although it is frequently difficult to get precise definitions, we have to persevere. We have to figure out what measure or range of measures has a bearing on these vague customer requirements. We should aim to provide an *operational definition* that removes all ambiguity by describing what something is and how to measure it, by translating vague concepts into procedures that everyone can use.

> **Q3.** *How might you measure good or bad service in a restaurant? Write down an operational definition of good service.*

2.4 Collecting Data

Care has to be taken in collecting data. Check sheets are the primary tool for gathering data. They are structured forms that make it easy to collect data. A well designed check sheet improves the reliability of the information collected. It also builds a picture of what is going on in the data, thus simplifying further analysis. An example of a check sheet is shown in Figure 2.4 in the debt recovery case study.

The following guidelines will help you make good check sheets and gather good data:

- Use *operational definitions*. Operational definitions remove ambiguity from check sheets, for they define what is to be measured and how it is to be measured.

■ Incorporate a *visual element* if possible. For example, suppose we are studying damage on cardboard cartons. The check sheet could consist of a picture of a carton, and the position of the defect can then be noted on this picture.

■ Include a section for recording *background information*, such as date, time, operator, and machine. Without this information it will be difficult to undertake a full analysis of the data later. For example, did the results differ from machine to machine?

■ *Test* your check sheet. Use the check sheet to collect and then analyse a small set of data. Did it work as well as you thought? Based on what you learned, improve the check sheet and then begin your data-collection effort.

■ *Design* it so that questions are answered from the top to the bottom of the page. Limit it to a single page, and allow extra room for comments.

■ Keep asking yourself, "Can it be improved?"

2.5 Flowcharts and Identifying Causes

Process flowcharts help us to understand a process by mapping out, step by step, in a diagrammatic form, a sequence of actions that take place in the process. They provide a means of arriving at a common understanding of what the current process does, an insight into where in the process poor quality, waste, delays, or inefficiencies might be occurring, and a good framework to show the effects of the problem solving effort.

The simplest type of flowchart is the *top-down flowchart*. It maps out the major steps of the process, concentrating on the essentially useful activities that take place. Having identified these major steps, we can then add more detail of what is involved in each step. If attention has been focused on a particular process for the purpose of problem solving, then a top-down flowchart is often all that is required to understand what is going on in that process. An example of a top-down flowchart is given in Figure 2.2 in the debt recovery case study.

Where the flow of a process is more complex or where more detail needs to be shown, other flowcharts may be necessary. *Decision flowcharts* split a process into different paths based on some decision being taken. *Deployment flowcharts* add further information about who or what department is responsible for each step of the process. *Work flowcharts* combine a plan of the physical layout of the process with indications of which route the work takes as it moves through the process. Scholtes, in *The Leader's Handbook* (McGraw-Hill, 1998), gives examples of these and other types of flowcharts.

Flowcharts are an invaluable aid in helping to identify potential causes of problems. A team working to solve a problem can use brainstorming methods to list causes. What this should involve is to record as many potential causes as possible. Team members should avoid evaluating whether these are likely to be real causes. That will come later. At this stage we want as many causes as possible, even if some of them eventually turn out to be unimportant. A flowchart will help the team focus on the various parts of the process in building up this list. A tool that has proven to be particularly helpful in setting out the potential causes in an organised way is the *cause and effect diagram*; it also called a *fishbone diagram* or an *Ishikawa diagram*. A well-constructed diagram, covering as many potential causes as possible, will help the team identify possible relationships between cause and effect, which will in turn help them to develop specific improvement strategies. An example of a cause and effect diagram is given in Figure 2.3 in the debt recovery case study.

2.6 Case Study: Debt Recovery Project

In the remainder of this chapter we shall follow through a problem solving process to show how improved results were obtained in the debt recovery project.

The *purpose, scope, and aims* of any problem solving project have to be clear at the outset. As stated in Section 1.3, the level of unrecovered debt has been a cause of concern for some time to the Albion Cleaning Company. The visits made by the general manager to the district offices have not been effective in reducing the overall level of debt. Even worse, they are costly and are contributing to a loss of morale, especially among the district managers. It is clear that this problem needs urgent attention:

- The high level of unrecovered debt is having a serious impact on the company's cash flow.
- Lost interest alone on the $1 million of unpaid invoices each month is not insignificant.
- The problem is believed to be leading to numerous customer complaints and to upset and lost customers.

An improvement team consisting of the general manager and other senior managers was set up to study this problem. The team members' aim was to determine the causes of the high level of unrecovered debt, to eliminate those causes, and to prevent them from recurring. They set the following goals:

1. In the short term, to produce results that are consistently above a target level for recovered debt of 80%, a figure already shown to be achievable
2. In the longer term, to achieve a level of debt recovery of 90%, a goal felt to be realistic.

At the start of the project it was clear that the senior management team did not have a clear picture of what was involved in generating accounts, issuing invoices, collecting payments, and recovering debt. To make sure that all members of the team had the same understanding of what was involved, a top-down flowchart of the debt recovery process was constructed. This flowchart is given in Figure 2.2. Note that an operational definition of a debt is given in Figure 2.2 as "cheque not received by due date." In this particular case, the due date was the 20th of the month following the date of the invoice.

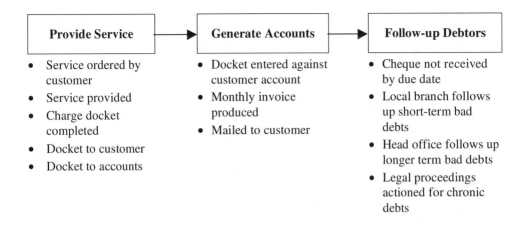

Provide Service	Generate Accounts	Follow-up Debtors
• Service ordered by customer • Service provided • Charge docket completed • Docket to customer • Docket to accounts	• Docket entered against customer account • Monthly invoice produced • Mailed to customer	• Cheque not received by due date • Local branch follows up short-term bad debts • Head office follows up longer term bad debts • Legal proceedings actioned for chronic debts

Figure 2.2   Top-down flowchart for debt recovery.

⚘ *Cause Analysis*

With the team now having a clear understanding of what was involved in issuing invoices, processing payments, and following up on debtors, it was then able to use brainstorming techniques to identify the potential causes of the problem. The team produced the cause and effect diagram in Figure 2.3, which lays out the process as a convergence of activities or causes that result in the final event or effect. Major cause/activity lines converge on the central line, and minor causes/activities that make up each major cause are plotted as short lines along the major lines. The diagram can "drill down" to as many levels as required, although three cause levels is usual, as shown here.

> **Q4.** *How are you going to use the information given in the cause and effect diagram of Figure 2.3?*

⚘ *Verifying Causes*

Overdue accounts have very many possible causes in the debt collection project. Which one should we address first? What are the most likely causes of the problem? The general manager thought that postal delays were the cause of the problem and said the company should use a courier company, instead of the postal service, to deliver the invoices to customers and to collect payments. How do we know this solution will work? We could try it and see, but this is what we do when we firefight a problem. If postal delays are not the cause of the problem, then we shall have put in place a solution that not only costs more but also does not lead to improvement. It may also be some months before this solution is shown to be ineffective.

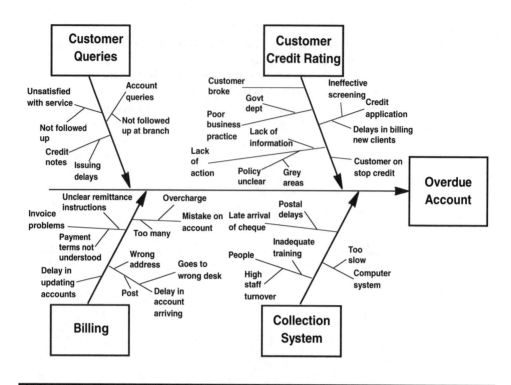

Figure 2.3   Cause and effect diagram for debt recovery project.

To avoid addressing just the symptoms of the problem, it is necessary to identify and verify the root causes. It is impractical to collect data on all the possible causes in the cause and effect diagram, so some prioritising is necessary. The team identified four possible causes that should be looked at first: *postal delays, account queries, mistakes on account,* and *government departments.* It decided first to collect some data on how long it took for customer payments to reach the Auckland office, where payments are processed.

The check sheet in Figure 2.4 was used to collect data on *postal delays.* The survey of mail was carried out on different days over a number of weeks. The results for one of the days are given in the check sheet. From the postmark, the number of days taken for the mail to reach the processing office in Auckland was recorded. The areas listed are in order from the north to the south of New Zealand.

A good picture of the delivery times is built up as the data are collected on the check sheet. A bar chart of these data is given in Figure 2.5. Similar results were obtained on the other days of the survey.

Q5. What do you conclude from the foregoing analysis?

DATE: *Thursday, 3 October 2002*

COLLECTED BY: *Jim Brown*

NUMBER OF PAYMENTS RECEIVED: *130*

Area Posted	Number of days to deliver payment				
	1	2	3	4	5+
Northland	|||				
Auckland	₥₥ ₥₥ ₥₥ ₥₥ ₥₥ ₥₥ ₥₥ ₥₥ ₥₥ ₥₥ ₥₥ ₥₥ ₥₥ ₥₥	₥₥				
Waikato	₥₥							
BOP / Hawkes Bay	₥₥ |||	||||			
Taranaki / Manuwatu		₥₥ ||||	||||		
Wellington	||	₥₥ |||					
South Island				||||			

Figure 2.4   Check sheet for postal delays.

The debt collection team concluded from the survey of incoming mail that postal delays were not a cause of the problem of overdue accounts. It revisited the cause and effect diagram and decided to collect information on *account queries*. The data obtained are plotted as a Pareto diagram in Figure 2.6. Pareto diagrams, which are discussed in more detail in Section 3.7, are basically bar charts with the main reasons ordered by magnitude.

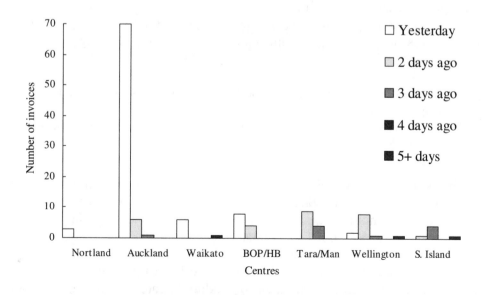

Figure 2.5   **Bar chart of postal delays (Thursday 3 October 2002).**

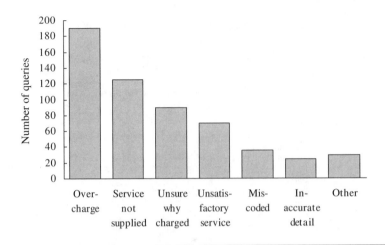

Figure 2.6   Account queries (November 2002).

Q6.   What do you conclude from Figure 2.6?

✎ *Solutions and Evaluating Results*

It was clear to the team that there was a very large number of account queries and that customers were not paying their invoices because they thought they were being overcharged or were not getting the service they thought they should be getting or were unsure what service they were getting. Various aspects of the service contracts were ambiguous and unclear, and the information given in the invoices was uninformative and incomplete. The actions taken by the team to remedy these difficulties were:

- To produce a brochure explaining the range of services offered and the cost of providing each type of service
- To redesign the invoice form so that it is clear to the customer what service has been provided and the cost of that service

The debt collection project team has now identified a cause of the problem that has been verified by data. It has taken action to remedy the situation. However, we cannot assume that the action taken will lead to an improvement in the results. We now need to evaluate the actions taken by comparing data on the process before and after improvement. For instance, in another project aimed at reducing the amount of damage on cardboard packaging, data clearly showed that the young, inexperienced forklift operators were causing more damage than the older, experienced operators. As a result it was decided to put

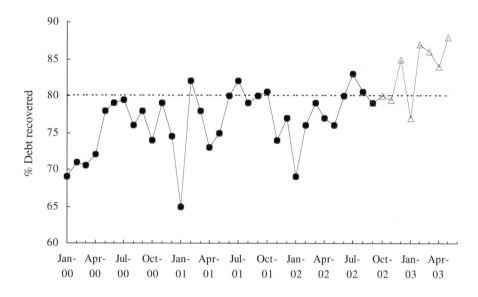

Figure 2.7 Debts collected within due month.

the young operators through an extensive retraining programme. Subsequent data revealed, however, that this solution had no effect. Further investigation led the project team to discover that the new operators were driving old forklifts that were no longer suitable. Once modifications were made to these forklifts, the differences between the operators were eliminated.

In the debt collection project the company continued to collect data on the percentage of debt recovered each month. The data up until May 2003 are shown in Figure 2.7.

> **Q7.** *Is there any evidence to suggest that the solutions put in place have been effective? In particular, what can you say about the January 2003 result?*

✍ *Standardizing and Planning for Future Improvements*

To ensure the solution is fixed in place, the team took the following actions:

- The brochure was issued to each salesperson and became the basis for negotiations with present and new customers.
- The new invoice form was adopted as the standard form; all previous forms were destroyed.
- A copy of the improvement story was sent to all district managers.

In reviewing what had been achieved, the team concluded that the improvements made had met the short-term goal of reducing the percentage of unrecovered debt to less than 20%. However, the team believed that still more could be done, and so it decided to continue working on this project. To reduce the percentage of unrecovered debt further, it looked again at the cause and effect diagram. As a result it decided to focus on those customers who were slow to pay. It suspected that this group included government departments.

Through understanding the nature of variability and the use of data and applying a scientific approach to problem solving, the senior management team had made significant improvements in just a few months. The percentage of unrecovered debt had been a concern for years, and all previous firefighting efforts had been unproductive.

2.7 Chapter Summary

This chapter has been concerned with how we can make informed decisions using data and a scientific approach to problem solving. Informed decisions include:

- The use of a scientific approach to problem solving rather than firefighting
- The reasons we need to collect data
- The need for clear operational definitions so that we are clear about the data we need to collect
- The use of well designed check sheets for collecting data
- The use of flowcharts to map out processes and cause and effect diagrams to systematically capture all the potential causes of a problem
- How the scientific approach can be used to improve a process by:
  - being clear about the purpose, scope, and aims
  - having a clear understanding of what is involved through the use of top-down flowcharts
  - identifying the potential causes by use of the cause and effect diagram
  - verifying the root causes of the problem by collecting more data
  - finding and evaluating solutions that address the root causes
  - standardizing and putting in place proposed solutions
  - planning for further improvements

2.8 Exercises

1. In the debt collection project, the team went through one cycle of the scientific approach.
   a. What lessons can the team learn from what it achieved?
   b. Can you think of any ways of improving what the team did?
2. A whiteware manufacturer is concerned about the occurrence of visible scratches, blemishes, and other surface imperfections on its refrigerators. Management first ordered a motivation programme to persuade workers of the importance of refrigerator appearance. Next, the quality of the paint was improved. Neither actions helped, so it was decided to add a special "touch-up" operation at the end of the production line to correct any surface defects.
   a. Critically comment on the actions taken by management.

    b. Can you suggest an alternative approach to the problem?

3. Explain the purposes of check sheets. Construct a check sheet to study the reasons for absenteeism in a factory.

4. A major hospital is concerned about the length of time required to get a patient from the emergency department to an inpatient bed. Significant delays appear to occur when beds are unavailable. What would be the first steps you would take to tackle this problem?

5. Mary Thompson, the general manager of an up-market hotel, is concerned by the number of customer complaints received over the last three months. What are the first steps Mary should take to address this issue?

6. A major car rental company prides itself on being customer focused and on providing first-class service. It now plans to collect data to verify this. What is the first step it should take?

7. A student is concerned that he failed a number of courses during his first year and as a consequence must retake the entire first year before being allowed to proceed to the second year. He wants to make sure this does not happen again. Brainstorm the problem and identify as many possible causes of failure as you can. Using your list, create a suitably organised cause and effect diagram.

Chapter 3

Looking at Data

3.1 Types of Data

At first sight, it is tempting to think that data are just numbers that describe some process or situation. However, consider the following three pieces of information that describe a particular company:

- It is an insurance company.
- Its financial security rating is AA.
- In the last financial year, its turnover was $756.3 million.

Each of these can be considered as a piece of "data," but only the third one is expressed numerically. The first two are examples of *qualitative* data, whereas in the third example the data is *quantitative*.

✵ Qualitative Data

Within the category of qualitative data, it is usual to distinguish between *attribute* (or *nominal*) data and *ranked* (or *ordinal*) data. Attribute data simply describes a characteristic (attribute) that some item or individual possesses, which either serves to distinguish one item from another, or classifies each item into one of various mutually exclusive (and usually exhaustive) categories. The first example above is attribute data that describes the activity of the company, and distinguishes it from other types of business such as manufacturing or distribution. Other examples of attribute data are:

- A player's number in a football team. Note that numbers here are only used conventionally. They could equally well be letters or the players' names, which are sometimes used as well as (or instead of) numbers.
- A list of the different types of mistakes made on a particular job.

- Reasons for absences from work. These could be categorised into *sickness, injury, leave with pay, leave without pay, incapacity of a close relative, other.* The final category is usually needed to cope with rare, infrequent reasons and with situations we may not have thought about in advance.
- Marital status, as on immigration forms at airports, using categories such as *never married, now married, separated, divorced, widowed.*
- Those where there are just two categories (the simplest type of attribute data). For instance, *defective* and *satisfactory, good* and *bad, pass* and *fail,* and *yes* and *no.*

The categories or labels used are often expressed as numbers, for example, 0 = male and 1 = female. This should not disguise the fact that such data are not truly numeric, meaning arithmetic is not appropriate. For example, it is not sensible to add up a set of such 0s and 1s to give an "average" sex. The only property possessed by attribute data is that of equivalence. For example, if two people each have value 0, then they are the same sex (male).

As soon as order is implied between individuals or categories, then the data are *ranked* (or *ordinal*). It may be that one category is preferred to or is better than or is higher than another category. In whatever way, the individuals or categories can be placed in an implicit order. Our earlier financial security example involves ranked data, in that the categories used (AAA, AA, A, BAA, etc.) are in decreasing order of financial security. In some cases, all the items or individuals forming a sample of data are ranked from first to last in some way, such as the results of a horse race, with few or no ties. More often, each item is placed into one of a number of ranked categories, such as the grades you get on a course (A$^+$, A, A$^-$, B$^+$, B,...), with items falling in the same category being seen as equivalent but as higher or lower than other categories.

This latter kind of data often arises in market research where respondents are asked to express their opinions about something on a 5-point scale, such as *strongly agree, agree, indifferent, disagree,* and *strongly disagree.* It is important to note that, although one category is higher or lower than another, it is not possible to infer how much higher or lower. In particular, the "difference" between *strongly agree* and *agree* may not be the same as that between *agree* and *indifferent,* in the same way that the difference (in marks) between the first- and second-placed students may not be the same as that between second- and third-placed.

When dealing with ranked data, only calculations involving relative position are meaningful. For example, it is possible to calculate an average of ranked data using the median but not the arithmetic mean. The median is defined in Section 3.9 and the arithmetic mean in Section 4.3.

✤ *Quantitative Data*

Once it is possible to say by how much one observation is greater than another, the scale of measurement is called *interval.* Order exists again, but the magnitude between values has meaning and provides extra information. Turnover of a company is interval data, as are the following examples:

- The time taken by each horse to complete a race
- The daily closing prices on the London Stock Exchange
- The number of beers consumed weekly by students at the University of Waikato
- Figures that describe the state of some process, such as the amount of an additive, the temperature of an oven, the rotational speed of a motor, or the weight of packages produced

It is possible to distinguish different types of quantitative data (for example, between *interval* and *ratio* data), but this will not be relevant here, because most simple statistical procedures are applicable to all forms of quantitative data.

Often only attribute data are available, but greater information is obtained with ranked and, especially, interval data. At all times strive to obtain interval data, although this may not always be possible. Avoid throwing away information by converting interval data to attribute data, for instance, when the lengths of some critical component are measured but the only information recorded is whether these lengths are within or outside some specification limits. They could all be within the limits, so the attribute data says nothing, whereas an important difference could be indicated by the interval data.

Q1. *List three further examples of attribute data and interval data.*

3.2   Presentation of Data

Particular attention should be given to the presentation of numerical results. Presenting data badly in tables or graphs can make any report ineffective. It can also mislead. Consider the following example.

Life Insurance Example

The data in Figure 3.1 gives the amount of ordinary life insurance (in millions of dollars) for each quarter in the years 1999 to 2003.

Quarter	Year				
	1999	2000	2001	2002	2003
1	3024	3867	4811	4298	6372
2	3156	4305	5914	5540	6837
3	3313	3673	4152	5203	6339
4	3860	4429	5217	5825	7640

Figure 3.1   Quarterly purchases of life insurance (in millions of dollars).

> **Q2.** *Do you find this table easy to read at a glance? List some of the things that make it hard to read that distract you from making a useful summary of the information contained in it.*

A number of suggestions for more effective presentation will now be considered. The books by A. S. C. Ehrenberg (*A Primer in Data Reduction,* John Wiley & Sons, 1982) and E. R. Tufte (*The Visual Display of Quantitative Information,* Graphics Press, 1983) are the basis for these suggestions.

3.3   Rounding Data

One major problem with reading numerical data is that it is difficult to conceptualise and mentally manipulate long numbers. A possible way of overcoming this difficulty is to round each number to *two effective digits*.

It is common practice to show percentages to three significant digits, such as 17.9% and 35.2%, but with numbers such as these it is difficult to mentally divide the smaller into the larger or even to subtract the smaller from the larger. Rounding to two effective digits gives 18% and 35%. It is now easy to see that one is just under twice as large as the other and that the difference between them is about 17 percentage points.

Consider index numbers such as 117.9, 135.2, 128.6, and 144.3. The initial 1's are not "effective," since they do not vary in these data. They are "dead" digits. The next two digits are the effective, or "busy," digits. Therefore, the numbers to two effective digits are 118, 135, 129, and 144. These numbers are still quite easy to manipulate because the initial 1's are the same. However, if the next index number is 93.3, then rounding to two effective digits would give 90, 120, 140, 130, and 140, which is far too great a discrepancy. Now it is better to use three effective digits, to give 93, 118, 135, 129, and 144. The basic idea is to use as few digits as possible, for ease of comprehension, while introducing as little distortion as possible by the rounding.

When there is wide variation in data, a variable standard for rounding can be used. For example, values such as 223.3 and 34.7 would be rounded to 220 and 35, respectively. This variable rounding does not really affect any comparisons between the numbers.

The information lost by dropping the other digits is usually negligible, yet the advantage in terms of communication is considerable. It is not suggested that the original data be discarded. It may be that the other digits are important, for instance, in monetary terms. The rounded data *are to be used only* in reporting the results in order to add clarity to the report.

3.4 Effective Tables

Data are very often presented in tabular form, such as in Figure 3.1. Several simple principles can be followed that will make a table easier to follow:

1. *Round the results to two effective digits,* as described in the previous section.
2. Use *averages* to indicate the overall structure of the table. Row and column averages can provide a focus to guide the eye when looking at the individual figures. Even when the pattern varies from row to row or from column to column, the row and column averages provide a useful focus to help see the nature of the differences.
3. Where possible, *order rows and columns by size.* This helps to identify patterns and exceptions to the data. This is not always possible, however, if the categories concerned have a natural order, such as the columns in Figure 3.1.
4. *Numbers are easier to read down than across.* This is because the leading digits in each number are then close to each other for direct comparison, with no other digits in between. Thus the table should be arranged with the categories whose comparison is of greatest interest down the left-hand side of the table.
5. Have a *simple layout* that is easy to read. Table layout should make it easy to compare the relevant figures (the two preceding guidelines are also based on this principle). The primary reason for the existence of the table is to convey information, not simply to impress the reader. In particular:
   a. Too many gridlines, widely spaced rows, and irregular spacing of rows and columns impede visual comparison and can be distracting.
   b. The intelligent use of white space in the layout can help convey the underlying groupings and emphasise any patterns in the data.
   c. Reducing the size of a table can also be helpful. It reduces the amount of eye movement required to scan the table.
   d. All tables should, of course, be clearly labelled.
6. A brief *verbal summary* helps to focus attention on the salient features of the table and consequently makes the whole presentation so much easier to read and understand.

Q3. *Using the guidelines just given, set out the data in Figure 3.1 in a more easily comprehensible table.*
Q4. *Write a sentence or two to summarise the information contained in these data.*

3.5 Graphs

Graphs serve essentially to convey information about relationships, shapes, relative sizes, and priorities rather than to present strict numerical information.

They help to make large sets of data coherent. Keep a few things in mind when creating graphs:

- If a complex story is to be illustrated graphically, then it is better to do so in a series of graphs, each of which illustrates a single element of the story.
- Graphs can reveal the data at several levels of detail, from the broad overview to the fine structure.
- It is important that all aspects of the graph be clearly labelled. The graph should have a heading that gives sufficient information to identify the original data. The axes should have titles, and the scales should be clearly marked.
- Overelaboration should be avoided in constructing graphs. The substance of the graph should be the centre of attraction, not the design graphic, the technology of graph production, or something else. Overelaboration often leads to a distortion of the information contained in the data.

In the following sections, some of the more commonly used graphs will be described and illustrated.

3.6   Bar Charts

The simplest form of graph is the bar chart, which can be used for attribute or ranked data. We have already seen an example of a bar chart in Figure 2.5, which depicted postal delivery times. A bar chart helps to compare the magnitude of various categories that are measured in some way. The most usual measure is the number of values occurring in each category, but other measures are possible. For example, the bar chart in Figure 2.5 could have been presented with just a single bar for each centre, where the size of the bar reflects the average delivery time from that centre.

Consider another example, relating to the issue of credit notes by the Alphington Depot as a result of errors in the supply of goods. The check sheet in Figure 3.2 records the reasons for the issue of 212 credit notes during a two-month period. A list of reasons is drawn up for the issue of credit notes; as each note is checked, the reason for its issue is tallied. A good picture of the patterns of credit note issue is built up.

Most graphs, including bar charts, are very easily created in Excel, as described in Section 9 in the *Introduction to Excel* on the CD-ROM. First the data in Figure 3.2 must be entered into a spreadsheet, as in Figure 3.3. Using the Chart Wizard, follow steps 1 to 4 selecting a column chart, with data in A2:B9 arranged by columns. The X and Y axis titles are input in step 3, and the legend and gridlines suppressed. Pressing *Finish* creates the required bar chart on the spreadsheet, as shown in Figure 3.3. Each of the reasons (categories) identified in Figure 3.2 is represented by a bar, whose height is equal to the number of times that reason occurred.

LOCATION: *Alphington*   DATE INSPECTED: *Nov 2003*

PERIOD COVERED: *Sept / Oct*   INSPECTOR: *Alf James*

NO OF INVOICES ISSUED IN PERIOD: *3797*

REASON	TALLY	
Damaged Goods	‖	2
Incorrect Goods	⅟⅟⅟ ⅟⅟⅟ ⅟⅟⅟ ⅟⅟⅟ ⅟⅟⅟ ⅟⅟⅟ |	31
Nondelivery	⅟⅟⅟ ⅟⅟⅟ ⅟⅟⅟ ⅟⅟⅟ ⅟⅟⅟⅟	24
Not Required	⅟⅟⅟	3
Out-of-Stock	⅟⅟⅟ ⅟⅟⅟ ⅟⅟⅟ |	16
Pricing Error	⅟⅟⅟ ⅟⅟⅟⅟⅟	108
Shortage	⅟⅟⅟ ⅟⅟⅟ ⅟⅟⅟ ⅟⅟⅟ ‖	22
Other	⅟⅟⅟ |	6
TOTAL		212

Figure 3.2   Check sheet for credit notes.

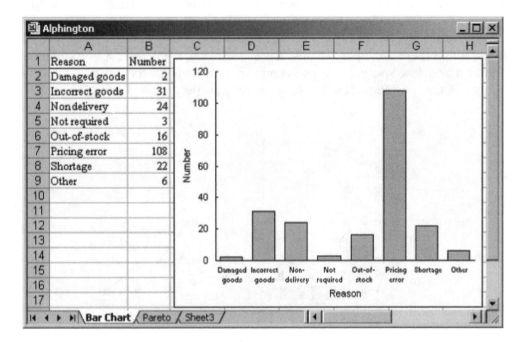

Figure 3.3   Bar chart of Alphington credit notes.

From Figure 3.3, it is quite clear that pricing errors are the most frequently occurring reason for the issue of credit notes. But what proportion of all credit notes is due to pricing errors? To make even a rough estimate of this proportion is not easy. This is because Figure 3.3 does not show the information in the most suitable form. When we are primarily interested in the proportion of the total that is accounted for by each category, the bar chart should be drawn to show the percentage occurrence of each category on the vertical axis. The only change necessary in Figure 3.3 is to replace the frequencies in column B of the spreadsheet by corresponding percentages. Alternatively, the data could be presented in the form of a *pie chart,* where the categories are represented by appropriately sized segments of a circle.

3.7 Pareto Diagrams

The Pareto diagram is a bar chart that ranks categories (items) by how important they are. Each block represents an item; the bigger the block, the more important it is. For example, when the items are problems, the Pareto diagram helps to identify which problems need immediate attention and which can be looked at later. While there are many things that cause a large problem, it is frequently found that a few specific factors account for the bulk of the observed occurrences of that problem. The Pareto principle can be depicted as in Figure 3.4.

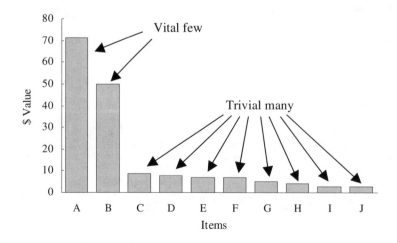

Figure 3.4 The Pareto diagram.

It has been shown many times in practice that a lot more progress can be made if problem solving efforts are concentrated on the main causes of a problem rather than on trying to tackle a bunch of lesser causes. Solving the biggest problem will usually result in the greatest improvement in the shortest amount of time. If attempts are made to solve all problems at the same time, then some progress will probably be made, but it will usually be slower and may even be counterproductive. Once the largest problems have been attacked, then attention can be directed to the lesser ones, if this is warranted in terms of time and money.

With the categories arranged in order of importance, it is also meaningful to plot on the diagram the cumulative values arising from each category. The cumulative value for a category is the sum of the values corresponding to the particular category and the categories to the left of that category in the Pareto diagram. The cumulative line is particularly useful when looking at the overall effect of making an improvement.

Credit Note Data

In the bar chart of the Alphington credit note data given in Figure 3.3, the different reasons, with the exception of *Other*, were plotted in alphabetical order. This is one possible ordering. But others could be chosen, for the order of categories does not matter. The Pareto diagram, setting out the categories in order of importance, is given in Figure 3.5, together with the cumulative line.

Because a Pareto diagram is no more than an ordered bar chart, the procedure in Excel is identical to that for the bar chart, provided that the data are

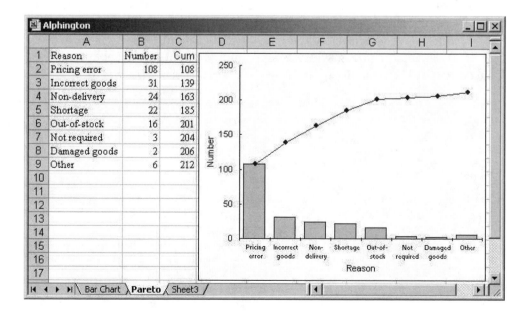

Figure 3.5   Pareto diagram of Alphington credit notes.

first arranged in descending order in column B. Since the *Other* category comprises a number of other reasons, it is usually placed last in the diagram. The cumulative line is obtained by choosing *Line-Column* from *Custom Types* in the Excel Chart Wizard; see Section 9 in the *Introduction to Excel* on the CD-ROM.

Q7. Explain what the diagram tells you.

The raw frequencies of occurrence may not be the best measure to plot on the Pareto diagram. If the value of each category is different, something more complex might be used. For example, looking at warranty failures of refrigerators, a broken plastic component costs much less to fix than a failed compressor. Thus the total costs of different categories of failure might be recorded, rather than simply recording the number of claims in each category.

✤ Evaluating Improvement

Pareto diagrams also enable improvement to be evaluated. Having identified potential major causes of a problem and taken corrective action, you can create another diagram to see how much improvement has taken place. If effective measures have been taken, then the order of items along the horizontal axis will usually change. In one problem solving exercise, a company was able to reduce the amount of damage occurring on the sides of their cardboard cartons by replacing hooked clamps on their forklift trucks with regular clamps. The Pareto diagrams in Figure 3.6 show very clearly the effect of this solution. With the cumulative lines drawn on the diagram, the overall effect of the improvement can be readily seen and measured. For instance, before improvement there were over 1600 damaged cartons, whereas after improvement this number had been reduced to just less than 1000.

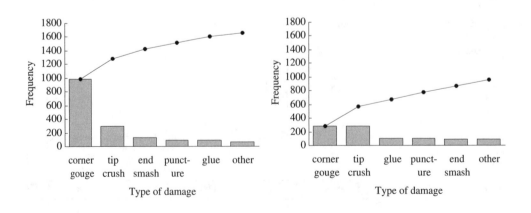

Figure 3.6 Pareto diagrams before and after improvement.

133.0	159.5	163.6	156.6	164.7	178.2	191.2	156.2
138.7	115.6	204.8	146.1	169.1	167.7	137.4	183.3
154.7	161.3	151.1	174.5	148.2	160.4	145.7	175.2

Figure 3.7   Sample data for stem and leaf plot.

Q8.   What was the main effect of the change? Did it affect all types of damage found on cartons?

3.8   Stem and Leaf Plots

When a large amount of interval data has been collected, it is important to get an overall picture of it. A useful graphical aid for looking at interval data is the stem and leaf plot.

The data are truncated to two effective digits:

■ The first effective digit is used to form the stem.
■ The second effective digit is used to form the leaves.

Other digits can be ignored, because they will make virtually no difference to the picture obtained.

Consider the data in Figure 3.7. The two effective digits are the tens and the units. Draw a line down the page, and list the stems on the left side, with one stem on each line, in ascending order. Put the leaves next to the appropriate stem, with one leaf for each measurement. Redraw the plot, rearranging the numbers on each stem in ascending order. The resulting stem and leaf plot for these data is shown in Figure 3.8. The data range from 115 to 204.

```
        Unordered                    Ordered
   11 | 5                       11 | 5
   12 |                         12 |
   13 | 3 8 7                   13 | 3 7 8
   14 | 6 8 5                   14 | 5 6 8
   15 | 9 6 6 4 1               15 | 1 4 6 6 9
   16 | 3 4 9 7 1 0             16 | 0 1 3 4 7 9
   17 | 8 4 5                   17 | 4 5 8
   18 | 3                       18 | 3
   19 | 1                       19 | 1
   20 | 4                       20 | 4
```

Figure 3.8   Stem and leaf plots.

Q9. *What does the stem and leaf plot tell you about these data?*

✍ *Use of Stem and Leaf Plots*

The overall shape of the stem and leaf plot provides valuable information with which to assess and compare data:

- Is the plot symmetric or skewed? It is easier to work with and compare symmetric shapes. With skewed data it is sometimes desirable to transform the data by taking, for example, logarithms or square roots before carrying out any detailed analysis.
- Does the plot have a single peak, or two or more peaks? Two or more peaks might indicate that the data are the mixed output from two or more different processes. Perhaps this is important. It does not necessarily give the answers, but it is pointing out a possibility.
- Are one or two data values very different from the others? We call such values *outliers*. They could indicate that something special was occurring in the process when these outliers were collected. On the other hand, such values could arise because some mistake had been made in collecting or recording the data.
- Are there clusters of data points, or does the plot drop off suddenly at either end? These may indicate the presence of a number of processes at work. Or it could be that items have been inspected and those at the extremes discarded.

✍ *Use of the* STiBstat *Add-In*

Drawing a stem and leaf plot in Excel can be difficult, for Excel has no built-in function that allows you to do this automatically. However, we have included some additional features that enable you to obtain stem and leaf plots (and other forms of display) easily. They can be found in the *STiBstat* add-in. This can be installed in Excel following the instructions given in Section 8 in the *Introduction to Excel* on the CD-ROM. The options available in the add-in are shown in Figure 3.9.

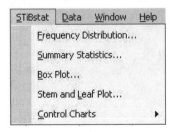

Figure 3.9   Options available in the *STiBstat* add-in.

To use the add-in to draw a stem and leaf plot, you must specify an array containing the data (the *Data range*) and a location in your spreadsheet where the plot will be written (the *Output range*). In addition, you may optionally specify a cell containing a *Variable label* (a name for your data) and a *Stem increment*. The stem increment is the difference between the lowest possible values in adjacent stems. For example, in the stem and leaf plot in Figure 3.8, the lowest possible value in the first stem is 110 and in the second stem is 120. The stem increment is therefore 10. If you do not specify a stem increment, one will be chosen automatically. You may then increase or decrease the number of stems by redrawing the plot and checking either *More stems* or *Less stems*.

For ease of interpretation, the plot specifies both the stem and leaf units. These are the numbers by which you must multiply the stated stem or leaf to get the actual value. For example, the first stem in Figure 3.8 is given as 11, but this actually represents the value 110, so the stem unit is 10. The leaf unit is always the stem unit divided by 10. In addition, if you choose to *Exclude outliers*, these will not be included in the plot itself, but the number of outliers present will be indicated below the stem and leaf plot.

Texona Savings Bank

The Texona Savings Bank has recently been reviewing the time it takes to process applications for a loan. The bank's target is to contact the applicant within three days of the initial application. William Hicks, the customer service manager, surveyed last month's loan records and extracted the data in Figure 3.10 on the number of days taken to process 37 loan applications.

The data have been entered into cells A1:A37 in an Excel worksheet called "Savings Bank." Clicking on the *Stem and Leaf Plot* option in the *STiBstat* add-in gives the dialogue box in Figure 3.11. The location of the plot in the worksheet will be determined by the cell given in the *Output range* box, namely, cell F1.

A stem and leaf plot of these data is given in Figure 3.12. Because the data only contain single digit values, there are no distinct leaves in this case, and zeros have been used. Any other character could have served, such as an ×, to simply show the shape of the distribution.

> **Q10.** *What percentage of loans is not meeting the standard of being processed within three days of the initial application?*

2	3	3	2	4	5	4	5	4	3	3	2	2
4	1	2	2	1	3	3	7	2	3	4	3	3
4	2	1	4	2	5	3	3	2	3	5		

Figure 3.10   Number of days to process loan applications.

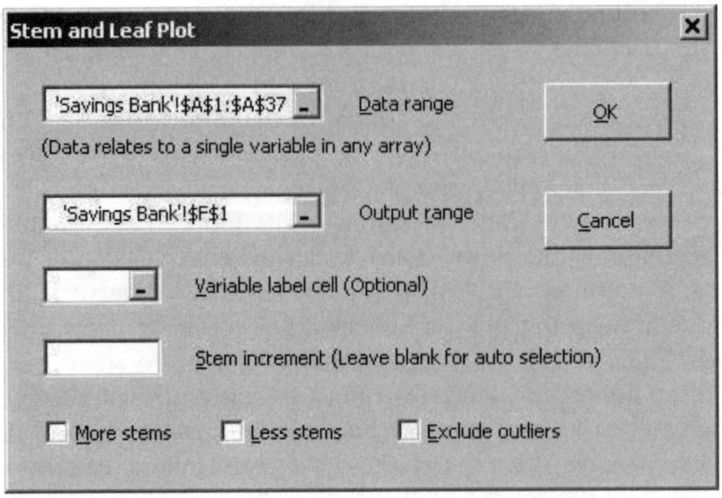

Figure 3.11   Stem and Leaf plot dialogue box.

3.9   Box Plots

The box plot is a useful graphical method for picturing the shape or distribution of a set of data. It is particularly useful to compare different groups of observations, by constructing a box plot for each group (on the same scale). A comparison of the plots will tell you about the symmetry of the distributions, as well as about the centre and spread. A box plot of the data in Figure 3.8, shown in Figure 3.13, was drawn using the *STiBstat* add-in. Selecting the *BoxPlot* option opens a dialogue box where you specify the range for your data, which can contain any number of variables in adjacent columns. Optionally, the first cell in each column can be a variable name, which will appear on the box plot. Specifying a range containing more than one column (variable) will produce a multiple box plot similar to that shown in Figure 3.16.

The box plot consists of a rectangle (*box*) divided by a horizontal line at the *median*. The median is a measure of the location (average) of a distribution and is the value that divides the data in half. In other words, half of the data

Frequency	Stem	Leaf
3	1	000
10	2	0000000000
12	3	000000000000
7	4	0000000
4	5	0000
0	6	
1	7	0

Unit	1	0.1
Stem increment	1	

Figure 3.12   Stem and Leaf plot for Loughborough Savings Bank.

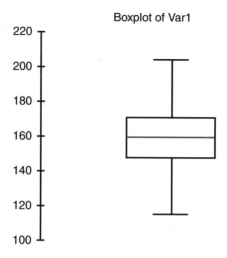

Figure 3.13 Typical box plot.

are below the median and half above. From Figure 3.13 you can see that the median is about 160. The ends of the box correspond to the upper and lower *quartiles*. The *lower quartile* is the value such that we have one-quarter of the data below it and three-quarters above it. Similarly, the *upper quartile* is the value such that we have three-quarters of the data below it and one-quarter above it. From Figure 3.13, the lower quartile is just less than 150 and the upper quartile just over 170. Joined to each end of the box are vertical lines (*whiskers*), which usually extend outward to the smallest and largest values in the data, in this case 115 and 204. Situations where the whiskers are drawn differently are discussed in the case study in Section 3.15.

The precise calculation of the median and quartiles will now be described.

✤ *Median and Quartiles*

Consider again the (ordered) stem and leaf plot in Figure 3.8. The 24 data values are set out in order of magnitude. The value in position 12 is 159, and the value in position 13 is 160. Twelve values are less than or equal to 159, and 12 values are greater than or equal to 160. To find the *median*, where we have an equal number of data values above and below it, we need to take the midpoint of the values in positions 12 and 13, i.e., 159.5.

In general, the median is given by the value in position $(n + 1)/2$, when the data are written in order of magnitude. For example, if $n = 23$, then the median is the value in position 12; 11 observations will be less than this value and 11 greater. If $n = 24$, then the median is the value in position 12.5, that is, halfway between the values in positions 12 and 13.

The *lower quartile* is given by the value in position $(n + 1)/4$. For our data (with $n = 24$) it is the value in position 6.25, namely, one-quarter of the way between the values in positions 6 and 7. Since these values are 146 and 148, respectively, the lower quartile is 146.5.

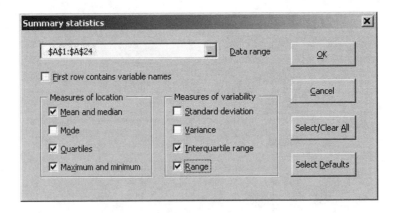

Figure 3.14   Summary statistics dialogue box.

The *upper quartile* is given by the value in position $3(n + 1)/4$. For our data it is the value in position 18.75, namely, three-quarters of the way between the values in positions 18 and 19. Since these values are 169 and 174, respectively, the upper quartile is 172.75.

The median and quartiles are available in Excel as functions *MEDIAN* and *QUARTILE*. However, the use of *QUARTILE* is not recommended, because it does not follow the definitions just given. Instead, all of these quantities can be readily obtained using the *Summary Statistics* option in the *STiBstat* add-in. In the dialogue box in Figure 3.14 the data are in cells A1:A24 and the required options chosen. The requested output is contained in an additional worksheet called *Output*, which is created automatically by *STiBstat*. This is shown in Figure 3.15.

The *range* of the data is the difference between the smallest value and the largest value. The *interquartile range,* that is, the difference between the lower and upper quartiles, is the range within which the middle half of the data lies.

Summary Statistics

	Var1
No. of observations	24
Mean	159.46
Median	159.50
Upper quartile	146.50
Lower quartile	172.75
Maximum	204.00
Minimum	115.00
Interquartile range	22.75
Range	89.00

Figure 3.15   Summary statistics for data in Figure 3.8.

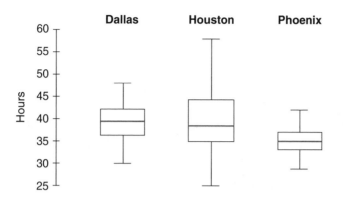

Figure 3.16 Box plots of mortgage application data.

📈 Texona Savings Bank

In addition to its target of three days for responding to loan applications, the Texona Savings Bank says that it will process a mortgage application within 48 hours of the application. William Hicks, the customer service manager, has collected the data on the number of hours taken to process mortgage applications at the main city branches in Dallas, Houston, and Phoenix. Fifty observations were collected at each branch. A box plot of the data is given in Figure 3.16.

> **Q11.** *What conclusions can you draw from these data as to whether the Texona Savings Bank is meeting its target in the three cities?*
>
> **Q12.** *Compare these data with regard to location, spread, and shape.*

3.10 Histograms

Figure 3.17 is a stem and leaf plot of the beads history data given in Figure 1.9. There are 425 data values and, as can be seen, the plot is fairly large. As more data become available, it is increasingly difficult to construct a stem and leaf plot of manageable size.

An alternative graph for plotting quantitative data is the *histogram*, which might be more appropriate in certain cases and particularly in those situations where there are large amounts of data. Histograms are readily obtained using the *Frequency Distribution* option in the *STiBstat* add-in. Figure 3.18 gives the dialogue box with data in cells B2:B426 and the *Histogram* option selected. The resulting histogram is given in Figure 3.19.

Frequency	Stem	Leaf
3	2	000
2	3	00
4	4	0000
15	5	000000000000000
22	6	0000000000000000000000
42	7	000000000000000000000000000000000000000000
40	8	0000000000000000000000000000000000000000
37	9	0000000000000000000000000000000000000
74	10	00000000000000000000000000000000000000000000000000000000000000000000000000
54	11	000000000000000000000000000000000000000000000000000000
48	12	000000000000000000000000000000000000000000000000
23	13	00000000000000000000000
19	14	0000000000000000000
24	15	000000000000000000000000
6	16	000000
7	17	0000000
1	18	0
2	19	00
0	20	
2	21	00

Unit	1	0.1
Stem increment	1	

Figure 3.17   Stem and leaf plot of the beads history data.

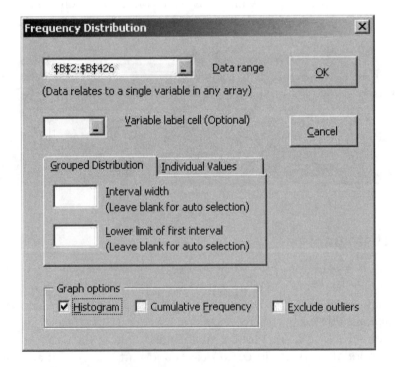

Figure 3.18   Frequency distribution dialogue box.

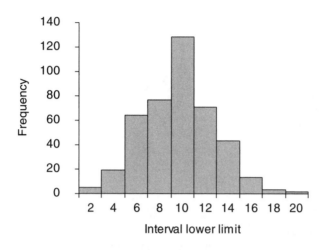

Figure 3.19   Histogram of the beads history data.

In constructing Figure 3.19, the interval width and the lower limit of the first interval have been chosen automatically. In this case the intervals correspond to pairs of stems in the stem and leaf plot in Figure 3.17. However, as can be seen in Figure 3.18, the user can specify the interval width or the lower limit or both. One possible disadvantage of this plot is that the exact number of values in each block is not given; the approximate number can, of course, be read off from the plot. However, if you want to show the exact numbers, then there is an option in Excel to do this. Click on the chart and from the *Chart* menu select *Chart Options*. Now choose *Data Labels* and then select *Value*. The number of values will then be written above each block.

Note that although the number of values in each block in a histogram is available, we do not know the individual values in a block. This information (to two effective digits) is available in a stem and leaf plot and is one of the advantages of the plot over a histogram.

Using the *Frequency Distribution* option will lead to histograms with equal interval width. In this case the number of items in each block is given by the height of the block. However, if you construct a histogram with unequal interval widths, then the number of items will be given by the *area* of the block. The height of each block is then the number of items divided by the width of the block.

3.11   Cumulative Frequency Plots

With Pareto diagrams it is useful to add a line representing the cumulative frequencies. An example was given in Figure 3.5, where the cumulative line clearly showed the overall effect of any improvement effort. A plot of cumulative frequencies or percentages is also useful when we have interval data.

As an example, consider again the stem and leaf plot of the beads history data given in Figure 3.17. From the first column of frequencies, the cumulative

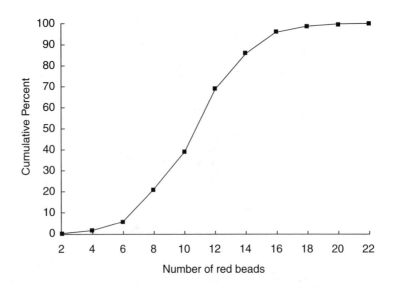

Figure 3.20   Cumulative frequency plot of the beads history data.

frequencies can be calculated and turned into percentages by dividing each cumulative frequency by the total frequency, 425, and then multiplying by 100. A plot of the cumulative percentages can readily be obtained using the *Frequency Distribution* option in the *STiBstat* add-in. It is only necessary to select the *Cumulative Frequency* graph option in Figure 3.18. The resulting plot is given in Figure 3.20. From this we can see, for instance, that the median is about 10.5, corresponding to the 50 cumulative percent point. Similarly, the upper and lower quartiles can be obtained as well as other percentile values.

📈 Shareholders Data

A financial consultant has collected data on the number of shares held by 1000 individual shareholders of a private company in the United States. A cumulative frequency plot of these data is given in Figure 3.21. The consultant is interested in the percentage of shareholders with more than 15,000 shares. From Figure 3.21 we can see, by extending a vertical line from 15,000 on the horizontal axis to the line on the graph and then across horizontally to the vertical axis, that about 95% of shareholders hold less than 15,000 shares. Consequently about 5% of share-holders hold more than 15,000 shares.

Q13. How many shareholders hold less than 10,000 shares?

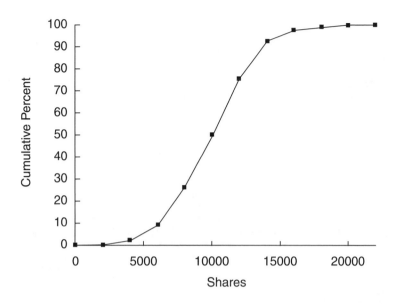

Figure 3.21   Cumulative plot of numbers of shareholders.

3.12 Run Charts

When data are collected from a process over some time period, it is often important to plot the data in the order in which they arose. This sort of graph is called a *run chart*. Plotting the time sequence may reveal, for instance, the existence of some trends that would not be indicated simply from an examination of a bar chart or stem and leaf plot. Also there may be changes in the average level at certain times, or changes in variability may be detected.

Various examples of run charts have been given in Chapter 1. An example is Figure 1.2, where the percentage of debt recovered is plotted over time. This chart was drawn in Excel by first entering the data into the spreadsheet, with the months in column A and the % debt recovered in column B. Selecting chart type *Line* at step 1 of the Chart Wizard, defining the data range (in columns) as A1:B33, and entering the required titles in step 3 gives the run chart of Figure 1.2, again having suppressed the chart legend and gridlines.

🏭 *Road Fatalities*

The numbers of people killed on New Zealand roads, per 100,000 population and per 10,000 vehicles, for each year from 1951 to 1987 are given in the run charts in Figure 3.22.

> **Q14.** *What are the important features of these charts? What questions would you ask?*

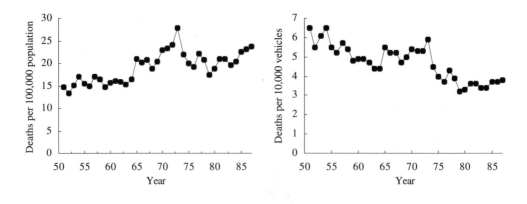

Figure 3.22   Fatal accident rates in New Zealand, 1951–1987.

3.13 Scatterplots

A scatterplot is used when we wish to study the relationship between two quantitative variables. One variable is plotted on the vertical axis (the *y*-axis) and the other variable on the horizontal axis (the *x*-axis). Each pair of values is given by a point on this graph. To create a scatterplot in Excel, first enter the data into two columns of a spreadsheet. Then using the Chart Wizard, select the *XY (Scatter)* option, and choose the type of scatterplot required. It is usual to have just the data points on the graph, without any joining lines. Finally, the graph can be customised by including axes and graph titles and modifying the axis scales if necessary.

📈 *Interest Rates and Housing Starts*

The Master Building Association believed that the number of building permits issued decreased dramatically during periods of high home loan interest rates. Figure 3.23 gives data for these two variables from 1971 until 1995. A scatterplot of these data is given in Figure 3.24.

> *Q15. What can you say about the relationship between interest rates and the number of building permits issued? Is there evidence to support the beliefs of the Master Building Association?*

Scatterplots are widely used in regression analysis, as will be seen in Chapters 8 and 9. As with other graphs, failure to plot the data in a scatterplot can lead you to overlook important features of, and relationships in, the data and to carry out inappropriate analyses.

	A	B	C	
		Home loan	Building	
1		interest rates (%)	permits (000s)	
2	Year			
3	1971	6.50	54.3	
4	1972	6.50	52.4	
5	1973	6.00	55.7	
6	1974	6.50	65.6	
7	1975	6.75	63.0	
8	1976	6.75	59.2	
9	1977	6.75	61.8	
10	1978	7.00	65.3	
11	1979	7.00	75.2	
12	1980	7.75	76.5	
13	1981	7.75	77.9	
14	1982	7.25	85.1	
15	1983	7.25	99.5	
16	1984	9.00	93.4	
17	1985	11.00	85.7	
18	1986	10.00	86.2	
19	1987	10.00	88.9	
20	1988	10.00	74.8	
21	1989	10.00	74.7	
22	1990	10.25	82.0	
23	1991	11.00	66.4	
24	1992	13.00	69.6	
25	1993	12.00	59.4	
26	1994	13.00	87.2	
27	1995	13.25	97.7	

Figure 3.23   Interest rates and building permits.

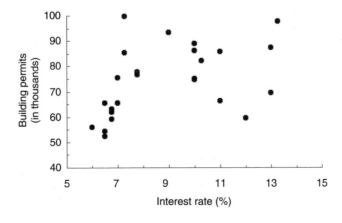

Figure 3.24   Scatterplot of interest rates and building permits.

Region	Target	Actual	Difference
Australia	3625	3715	+90
Japan	4675	4725	+50
UK	3000	2800	-200
USA	3200	3130	-70
Total	14500	14370	-130

Figure 3.25   Sales (in thousands of dollars) by region.

3.14   Stratification

We can often be misled when we collect data in aggregate. Important effects can be exposed by dividing data into subgroups, which can reveal a great deal of structure that may otherwise be unknown. Tracing information back to its sources is called *stratification*.

For example, a company buys raw material from two different suppliers. Stratification would involve breaking down the data into two groups, one for each supplier, and analysing them separately. A run chart for each supplier would allow a comparison of performance over time, while a stem and leaf plot would allow their performance against some standard to be compared.

Export Sales

The export manager is reviewing results for the September quarter and finds that actual sales are $130,000 down on the target figure of $14,500,000. She knows that the sales figures are never exactly on target, but $130,000 seems too large to ignore. She decides to investigate the source of the discrepancy and examines the sales data for the various sales regions. The information is shown in Figure 3.25.

Clearly the problem region is the United Kingdom. Should she sack the U.K. sales manager? The export manager decides to look at the performance of the four sales representatives in the United Kingdom. The results are given in Figure 3.26.

Sales Rep	Target	Actual	Difference
Henderson	750	780	+30
Smythe	800	550	-250
Whitaker	790	840	+50
Thomas	660	630	-30
Total	3000	2800	-200

Figure 3.26   U.K. sales (in thousands of dollars) by sales rep.

Outlet	Target	Actual	Difference
Hampshire	140	65	-75
Kent	110	70	-40
Sussex	105	60	-45
Dorset	130	65	-65
Devon	205	150	-55
Cornwall	110	140	+30
Total	800	550	-250

Figure 3.27   Smythe's sales (in thousands of dollars) by outlet.

Smythe is obviously the problem. Let's get rid of him! Not satisfied, the export manager calls for information on the business that Smythe has done with each of the six outlets he is responsible for. This produces the data on Smythe's outlet sales given in Figure 3.27.

There are no real clues here as to why Smythe's sales are below target. Perhaps the export manager should get rid of Smythe after all. However, Smythe has been with the company for many years and has been a good sales rep. Instead she decides to look at Smythe's sales by product line. This produces the data in Figure 3.28.

Now the real cause of the problem appears to have been identified. Ice Chill does not appear to sell within the United Kingdom. Should the product be withdrawn from the United Kingdom, or should an attempt be made to find out why it is not selling? That is another issue, but before any action is taken it would be wise to check whether the other U.K. sales reps seem to have the same problems with Ice Chill.

Tracing information back to its sources, in the manner of this example, is an important use of the stratification principle.

Moisture Content of Coke

The results on the percentage moisture content of coke (the fuel, not the drink!) produced at a coking plant taken daily over a 24-day period commencing on June 24 are given in Figure 3.29. The run chart for the data is given in

Product Line	Target	Actual	Difference
Export Lite	70	80	+10
Ice Chill	430	160	-270
Black Byte	250	250	0
Golden Glow	50	60	+10
Total	800	550	-250

Figure 3.28   Smythe's sales (in thousands of dollars) by product line.

Day	Moisture (%)	Coal Source	Day	Moisture (%)	Coal Source
24/6	9.32	A	6/7	9.37	B
25/6	9.78	C	7/7	9.42	B
26/6	9.38	B	8/7	9.34	A
27/6	9.44	B	9/7	9.78	C
28/6	9.76	C	10/7	9.44	B
29/6	9.30	A	11/7	9.19	A
30/6	9.33	A	12/7	9.32	A
1/7	9.82	C	13/7	9.72	C
2/7	9.35	A	14/7	9.45	B
3/7	9.78	C	15/7	9.68	C
4/7	9.78	C	16/7	9.36	B
5/7	9.46	B	17/7	9.34	A

Figure 3.29   Coke moisture content over a 24-day period.

Figure 3.30a. Important features of the data are clearly seen in this chart. For instance, the data are centred on 9.5%, and most of it lies within 0.3% of the centre. No obvious trend patterns over time can be seen. The source of the coal used in the coke production is also recorded in Figure 3.29. If this information is used to stratify the data in the run chart, then a completely different picture emerges, as can be seen in the run chart in Figure 3.30b.

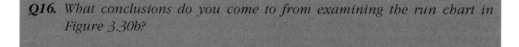

Q16. What conclusions do you come to from examining the run chart in Figure 3.30b?

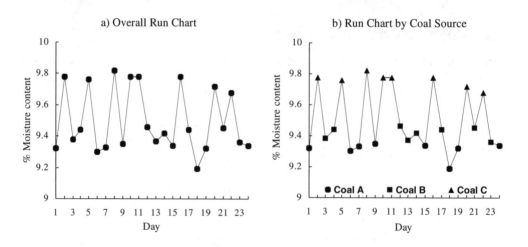

Figure 3.30   Run charts of coke moisture content.

13.15 Case Study: Aerofoil Effectiveness

The distribution centre of a large New Zealand manufacturing company was considering whether to fit streamlining aerofoils to their large trucks. The aerofoils are meant to give improved fuel consumption, but they cost $1000 each. In order to examine their cost effectiveness, the company decided to try an aerofoil on one of their trucks. They chose a "B" train truck and trailer unit that made a daily run from Auckland to towns in the central North Island, a round-trip journey that usually varied between 300 and 400 km.

The trial started on June 29 without the aerofoil. The aerofoil was fitted on August 15, removed on October 16, and refitted on November 22. The trial finished on December 14. A run chart of the fuel consumption is given in Figure 3.31.

> *Q17. What conclusions can you draw from this chart?*
> *Q18. Are there any surprising results? If so, what would you do about them?*
> *Q19. What other data would you like to see or to calculate?*

A run chart is very good for identifying patterns and trends over time. In this example, however, it is not clear from the run chart whether the use of the aerofoil has been particularly effective or not. Looking at data in different ways may give different insights as to what is actually going on. It may be useful to look at the distributions of the data when the aerofoil is both on and off. A stem and leaf plot for the aerofoil data is given in Figure 3.32. In this case since the

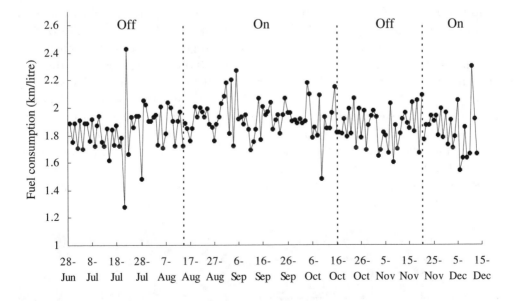

Figure 3.31   Daily fuel consumption.

ON		OFF
Leaf	Stem	Leaf
	12	8
	13	
	13	
	14	
8	14	8
4	15	
	15	
33	16	02
669	16	567799
123	17	001112222233
56677889	17	556889
00112444	18	011112234
555566677888999	18	556777799999
0001112223333344	19	000012223334444
555566667799	19	5678899
01134	20	012344
57789	20	5579
0	21	
588	21	
0	22	
7	22	
0	23	
	23	
	24	3

Figure 3.32   Fuel consumption (km/litre) with aerofoil on and off.

data has been stratified into two groups (namely, on and off), a *back-to-back* stem and leaf plot is useful for comparative purposes. Figure 3.32 was obtained by first constructing stem and leaf plots for both sets of data using the option in the *STiBstat* add-in and then cutting and pasting one plot onto the other. You will also notice from Figure 3.32 that each stem (with the exception of the first and the last) occurs twice. The lower leaves (0–4) have been placed on a separate line from the upper leaves (5–9) to give greater spread, and hence definition, to the stem and leaf plot. There are two possible ways in which the stems can be "split": either into two sets of five (as here) or into five sets of two. In the latter case, the leaves 0 and 1 would be on one row, 2 and 3 on the next, and so on. Whether it is necessary to split stems, and if so how, depends on the amount of variability in the data. Splitting is done automatically in *STiBstat* unless you chose to specify the stem increment.

Q20. *What do you conclude from this plot?*
Q21. *How might you summarise the information contained in this plot?*

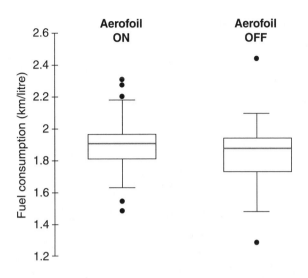

Figure 3.33   Box plot of aerofoil data.

The box plot is shown in Figure 3.33, and, again, it is particularly useful when comparing two distributions. Notice in this plot that some dots appear beyond the ends of the whiskers. The dots indicate that these data points are *possible outliers*, in the sense that they are some way from the middle half of the data (the box). For instance, when the aerofoil is off there is a value of 2.46 km/litre, which is a long way from the upper quartile of 1.97 km/litre. Extending the whisker to the largest value would distort the plot, because it would give the impression that one-quarter of the data were fairly evenly spread between the upper quartile of 1.97 km/litre and 2.46 km/litre, which is not the case. The rule adopted in constructing the box plot is that a dot represents any point that is more than 1.5 times the box size (the interquartile range) above the upper quartile or below the lower quartile. If a value is even further away from the rest of the data, it is a *probable outlier* (sometimes called an *extreme*). The definition of a probable outlier is that it is more than three times the box size away from the relevant quartile. Probable outliers are shown on the box plot by a small circle rather than a dot.

Q22. What conclusion would you draw from the box plot?
Q23. Summarise your conclusions from all the preceding plots. What other analyses might you wish to carry out?
Q24. When you examine the run chart in Figure 3.31 it is apparent that the two outliers when the aerofoil is off are on consecutive days (21 and 22 July). One of these is high and the other is low. What might this suggest about the data collection on these two days?

3.16 Chapter Summary

In this chapter we looked at the different types of data and how they can be presented to convey information in the most effective way. We examined:

- The distinction between qualitative and quantitative data:
  - Qualitative data involve attributes or categories that may be ordered or unordered.
  - Quantitative data arise where the magnitude of differences between values has meaning.
- The importance of presenting data in tables or graphs so that it is easy to understand.
- Some guiding principles for setting out data in tables.
- The graphical presentation of data by:
  - Bar charts and Pareto diagrams for showing patterns in qualitative data
  - Stem and leaf plots, box plots, and histograms to highlight the shape and features of quantitative data
  - Run charts to reveal patterns or trends of quantitative data over time
  - Scatterplots to show the relationship between two quantitative sets of data
- Stratifying data, whereby it is disaggregated in different ways to reveal hidden patterns.

3.17 Exercises

1. A timber processing company receives logs from two suppliers. The logs are dried in two different kilns, and the company records the moisture content, in time order, of the logs coming from the two different kilns and suppliers, as given in Figure 3.34. The moisture content data are plotted in Figure 3.35.

Moisture	Kiln	Supplier	Moisture	Kiln	Supplier	Moisture	Kiln	Supplier	Moisture	Kiln	Supplier
14.1	1	1	21.5	1	2	23.0	1	1	26.1	1	2
17.8	2	1	14.6	2	2	13.2	2	1	17.1	2	2
16.0	1	2	14.7	1	1	18.5	1	2	18.7	1	1
10.9	2	2	15.3	2	1	11.6	2	2	19.1	2	1
21.4	1	1	17.4	1	2	17.9	1	1	21.8	1	2
16.3	2	1	12.8	2	2	14.3	2	1	16.6	2	2
20.4	1	2	24.1	1	1	19.5	1	2	19.9	1	1
12.5	2	2	14.4	2	1	14.6	2	2	11.5	2	1
18.9	1	1	23.3	1	2	22.4	1	1	17.1	1	2
14.8	2	1	18.2	2	2	15.0	2	1	14.4	2	2

Figure 3.34   Moisture content of logs.

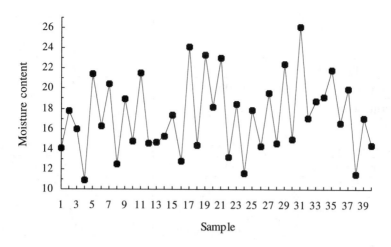

Sample

Figure 3.35   Moisture content of logs.

    a. What does the plot in Figure 3.35 tell you?

    b. Now plot the data using the information on the different kilns and suppliers. What do you now conclude from your plot?

2. One of the major measures of the quality of service provided by any organisation is the speed with which it responds to customer complaints. A large carpet company was receiving complaints about the installation of carpets. During a 6-month period, the number of days between the receipt and the resolution of a complaint is given by the data in Figure 3.36.

    a. Construct a stem and leaf plot of these data.

    b. Write a short paragraph that summarises the service time for resolution of these complaints.

    c. What actions should be taken to further study the process of resolving complaints?

3. The data in Figure 3.37 give the quality costs for a manufacturing company over the last five months of 2004.

    a. Set out these data in a table or tables that are easier to interpret. It may not be appropriate to present all of the data, so consider carefully which aspects of the data to present.

    b. Write a short paragraph to summarise the main features exhibited in your table(s).

4. Wilson Electronics wishes to source one of its major components from one of two suppliers. The choice has been narrowed to these two suppliers

68	33	23	20	26	36	22	30
52	4	27	5	10	13	14	1
25	26	29	28	29	32	4	12
5	26	31	35	61	29		

Figure 3.36   Number of days taken to resolve complaints.

Prevention Costs:					
Quality audit	$270	$270	$270	$270	$270
Cost of training	$1,250	$2,300	$600	$350	$600
Improvement projects	$1,000	$2,000	$400	$400	$600
Certification	$537	$537	$537	$537	$537
Appraisal Costs:					
Inspection wages	$10,000	$10,000	$7,400	$5,000	$6,600
Audit final inspection	$3,600	$3,600	$3,200	$2,400	$2,200
Calibration costs	$250	$250	$250	$250	$250
Specification changes	$500	$300	$300	$300	$300
	Aug-04	Sept-04	Oct-04	Nov-04	Dec-04
Internal Failures:					
Running rejects	$61,100	$43,176	$96,600	$41,064	$46,526
Inplant rejects	$8,500	$25,000	$19,660	$9,700	$15,000
Rework	$2,400	$2,000	$3,600	$4,400	$2,000
Downtime	$17,480	$10,400	$7,500	$500	$9,000
External failures:					
Customer credits	$20,000	$20,000	$16,567	$12,000	$2,900
Warranty claims	$2,500	$10,000	$7,000	$0	$6,250
Totals:					
Prevention	$3,057	$5,107	$1,807	$1,557	$2,007
Appraisal	$14,350	$14,150	$11,150	$7,950	$9,350
Internal failures	$89,480	$80,576	$127,360	$55,664	$72,526
External failures	$22,500	$30,000	$23,567	$12,000	$9,150

Figure 3.37   Quality costs.

because their components were about equal in all respects and far superior to their competitors in product quality. Bill Edwards, the supply manager, decided to look at the two suppliers' performance over the past 40 weeks. Each supplier has delivered one shipment every week. The days ahead (negative value) or behind the scheduled delivery date for each of the two suppliers, LCP and FEC, are given in Figure 3.38.

    a. Construct stem and leaf plots of the days after schedule for each of the two suppliers. Which supplier should Wilson Electronics choose, and why?

    b. Now construct run charts for each supplier, plotting both charts on the same graph. Which supplier should Wilson Electronics choose, and why?

5. The decline in the value of the British pound (£) between 1925 and 1975 is depicted in Figure 3.39.

    a. Why is this graph misleading?

    b. How would you plot these data?

Week	1	2	3	4	5	6	7	8	9	10
LCP	3.5	2.0	1.5	3.0	2.0	1.0	2.5	1.0	4.0	3.0
FEC	0.5	1.0	0.5	1.5	1.0	0.5	1.0	0.5	1.5	0.0
Week	11	12	13	14	15	16	17	18	19	20
LCP	3.5	2.0	2.5	4.0	3.0	3.5	3.5	2.5	1.5	1.5
FEC	0.5	1.0	1.5	1.0	1.5	0.0	1.0	0.5	1.0	2.0
Week	21	22	23	24	25	26	27	28	29	30
LCP	2.0	1.5	1.0	1.0	1.5	1.0	0.0	0.5	0.0	0.0
FEC	1.0	0.5	1.5	0.5	1.0	0.5	1.5	2.0	1.5	1.0
Week	31	32	33	34	35	36	37	38	39	40
LCP	1.0	0.5	0.0	0.5	-0.5	0.0	-0.5	0.0	0.5	0.0
FEC	1.5	1.0	0.5	1.0	0.5	1.0	1.5	0.5	1.5	1.0

Figure 3.38   Days after schedule.

6. The data in Figure 3.40 represent the number of daily calls received at a 0800 (toll free) number of a large Australasian airline over a period of 20 consecutive workdays (Monday to Friday). Write a brief report to the marketing manager in charge of the airline's customer service concerning traffic on the 0800 number.

7. Joe, the general manager of a mail order company, was concerned about the nonconformances occurring at its main customer invoicing centre. Joe decided to collect some data on the type and frequency of nonconformances and to carry out a Pareto analysis. Over a period of one month the data in Figure 3.41 were obtained.

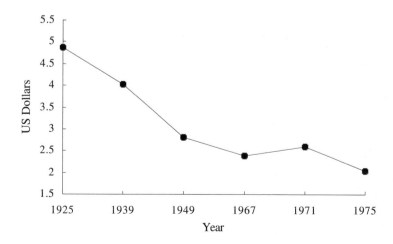

Figure 3.39   U.K.£–U.S.$ exchange rate, 1925–1975.

Week	Day	Number of calls	Week	Day	Number of calls
1	Mon	1060	3	Mon	1603
	Tues	1370		Tues	1256
	Wed	1087		Wed	1075
	Thur	1135		Thur	1187
	Fri	1805		Fri	1060
2	Mon	1234	4	Mon	1004
	Tues	1105		Tues	985
	Wed	1168		Wed	1618
	Thur	1235		Thur	1369
	Fri	1174		Fri	1353

Figure 3.40   Customer calls.

    a. What is the purpose of a Pareto analysis?
    b. Two Pareto diagrams could be constructed from the given data. What are they, and what are their relative merits?
    c. Construct both diagrams.
    d. Where should Joe concentrate improvement efforts?

	Type of nonconformance	Frequency	Cost per nonconformance ($)
1	Customer received goods but no invoiced processed	52	6.25
2	The invoice had no purchase number	25	0.72
3	Customer payment received but incorrectly input into the computer	23	0.85
4	Invoice had the wrong address	176	0.75
5	Customer name spelled incorrectly	202	0.40
6	Final demand sent to a customer who had already paid	8	1.05
7	Wrong description on invoice	35	0.75
8	Customer payment recorded on the wrong account	6	0.95
9	Invoice held up for more than three days in the sales department	111	1.50
10	Invoice sent to customer before the goods were	12	0.65

Figure 3.41   Nonconformance in customer invoices.

43	38	56	42	54	73	42	53
62	46	37	49	40	34	47	41
34	63	39	36	32	49	43	59

Figure 3.42   Times to fix breakdowns.

    e. Joe is considering firing the input operator because of all the noncon-
formances due to input errors. Explain why this might be an inappropriate action to take.

8. A machine shop manager wishes to study the time it takes an assembler to
complete a given small subassembly. Measurements, in minutes, are made
at consecutive half-hour intervals.
    a. Explain why a run chart might be an appropriate way of displaying these
data?
    b. What graphical method would you use if you were interested in looking
at the shape of the data?

9. For domestic customers, the Mars Electricity Company (MEC) rebates the
$20 monthly line charge fee if a breakdown in supply is not repaired within
48 hours. During one hot weekend the hours taken to fix breakdowns in
one city are as given in Figure 3.42.
    a. Construct a stem and leaf plot of these data.
    b. What proportion of time is MEC paying the rebate?
    c. Calculate the median time taken to carry out a repair.
    d. Give two actions that MEC could take to try and improve their breakdown
repair service.

10. The amount of space available per student at a university campus was
depicted as shown in Figure 3.43.
    a. Why is this graph misleading?
    b. How would you plot these data?

11. The manager of a local consumer association is concerned about reports
that similar generic medicines are being sold at widely differing prices at
local pharmacists. She collected the prices shown in Figure 3.44 of one
particular medicine at 30 different pharmacies.
    a. Construct a stem and leaf plot of these data. What does the plot tell us
about these data?

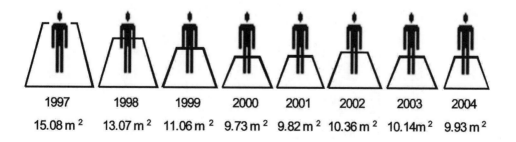

1997	1998	1999	2000	2001	2002	2003	2004
$15.08 \, m^2$	$13.07 \, m^2$	$11.06 \, m^2$	$9.73 \, m^2$	$9.82 \, m^2$	$10.36 \, m^2$	$10.14 \, m^2$	$9.93 \, m^2$

Figure 3.43   Space per student.

7.13	8.65	6.39	9.38	6.09	5.98
10.80	7.69	5.35	7.09	9.78	7.19
7.23	10.93	8.43	7.95	10.26	8.37
7.59	7.45	7.86	8.14	9.19	7.59
6.25	5.71	8.69	6.24	8.03	6.92

Figure 3.44   Prices of a medicine at different pharmacies.

     b. All the pharmacists buy the product from the same wholesaler at $4.50. Given that the manager considers a markup of more than 100% to be unfair to the consumer, what proportion of the pharmacies sampled will not meet with her approval?

12. A mechanical part that is machined on a lathe has recently been found to have defective hole diameters, which disrupt the assembly work that

Date	Alloy	Hole Diameter (0.0001mm)				
Sept 14	Al/Mn	7	24	24	24	25
15	Al/Mn	17	37	28	16	26
16	Al/Mn	12	22	40	36	34
17	Al/Mn	52	35	29	36	24
19	Al/Mn	28	28	34	29	48
20	Al/Mn	39	27	48	32	25
21	Al/Mn	36	21	31	22	28
22	Al/Mn	5	33	15	26	42
23	Al/Mn	50	34	37	27	34
24	Al/Mn	21	17	20	25	16
26	Al/Mn	34	18	29	43	24
27	Al/Mn	18	35	26	23	17
28	Al/Mn	10	28	19	26	21
29	Al/Mn	21	23	35	28	38
30	Al/Mn	27	41	15	22	23
Oct 3	Al/Hg	37	19	39	21	38
4	Al/Hg	37	46	22	26	25
5	Al/Hg	13	32	35	56	45
6	Al/Hg	9	51	25	37	39
7	Al/Hg	14	27	34	37	52
8	Al/Hg	30	51	34	36	28
10	Al/Hg	54	31	35	29	25
11	Al/Hg	45	21	38	38	31
12	Al/Hg	19	31	27	25	38
13	Al/Hg	25	45	41	36	43
14	Al/Hg	30	24	44	48	38
15	Al/Hg	64	32	32	42	42
17	Al/Hg	8	58	65	33	39
18	Al/Hg	38	37	50	37	33
19	Al/Hg	64	38	47	49	41

Figure 3.45   Hole diameters in an alloy.

Category	Annual loss ($)
Downtime	38,000
Testing costs	20,000
Rejected paper	560,000
Odd lot	79,000
Excess inspection	28,000
Customer complaints	125,000
High material costs	67,000

Figure 3.46   Quality losses at a paper mill.

follows. Data have been collected over a 30-day period, where five parts have been taken from the production process at regular intervals on each day. The data are given in Figure 3.45. Note that the alloy used was changed at the end of September.

   a. Construct two stem and leaf plots, one for the September data and the other for the October data. What can you say about the data from these plots?

   b. Construct box plots for each month. What conclusions do you draw from these plots?

13. Draw a graph to display the data on quality losses in a paper mill given in Figure 3.46. What conclusions do you draw from your graph?

Chapter 4

Modelling Data

4.1 Introduction

To think statistically requires an understanding of variation. As we saw in the first chapter, managers have to draw conclusions and make decisions in a variable and unpredictable world. We saw also, in Chapters 2 and 3, how data are important for effective decision making and that it is important for data to be appropriately tabulated and displayed in order to give meaningful insights into a problem. In this chapter we begin to look beyond the data to try and understand more about the situation that gave rise to the data. In particular, we will look at how data can be represented by simple statistical models.

A model is essentially an abstraction or conceptualisation of some situation that helps us to understand why we observe what we do. What is it that lies behind the data we have collected to cause them to be like they are? Once we have a model that adequately represents the data, we can ask "What if?" questions. For example, suppose we collect data from a doctor's surgery (office) on patient waiting and consultation times. Once we have postulated an appropriate model to describe, for example, the variation in consultation times, it can be used to estimate such things as the effect of a reduction in this variation or the provision of an extra doctor.

We will begin by looking at three different situations to motivate our consideration of how and why models of data are useful in decision making.

🏭 Debt Recovery

Consider again the debt recovery problem introduced in Chapter 1 and examined in more detail in Chapter 2. A run chart of 33 months' data on the percentage of invoices paid within the due month was given in Figure 1.2. The data concerned are given in Figure 4.1, with a stem and leaf plot in Figure 4.2.

Year	Jan	Feb	Mar	Apr	May	Jun	Jul	Aug	Sep	Oct	Nov	Dec
2000	69.0	71.0	70.5	72.0	78.0	79.0	79.5	76.0	78.0	74.0	79.0	74.5
2001	65.0	82.0	78.0	73.0	75.0	80.0	82.0	79.0	80.0	80.5	74.0	77.0
2002	69.0	76.0	79.0	77.0	76.0	80.0	83.0	80.5	79.0			

Figure 4.1   Debt recovery data: percentage of invoices paid within the due month.

Q1.   *What can you see from the stem and leaf plot in Figure 4.2 that is not easily seen in the run chart in Figure 1.2 or the data in Figure 4.1?*

Q2.   *How would you describe the variability in the percentage of invoices paid within the due month?*

📈 *Foodstore Supermarkets*

Foodstore, a national U.K. supermarket chain, has a policy of reducing by 40% the price of any fresh food items that are within two days of their sell-by date. Reduced items have a sticker covering the bar code, and the checkout operator must remove this sticker and enter the reduced price before scanning the item. To monitor the sales of reduced price items, the manager of the Hamilton store records the total saving obtained from buying reduced price items by a sample

Frequency	Stem	Leaf
1	6	5
0	6	
2	6	99
2	7	01
2	7	23
4	7	4445
5	7	66677
9	7	888999999
5	8	00000
3	8	223

Unit	10	1
Stem increment	2	

Figure 4.2   Stem and leaf plot of debt recovery data.

Frequency	Stem	Leaf
7	0	0000013
6	0	566999
1	1	3
1	1	9
3	2	144
7	2	5678899
9	3	111233444
4	3	5689
11	4	12223333444
7	4	5788888
4	5	0134

Unit	1	0.1
Stem increment		0.5

Figure 4.3   Savings per customer on reduced price items.

of 60 customers on the previous Saturday. A stem and leaf plot of savings (dollars per customer) is shown in Figure 4.3.

> **Q3.** *Is there anything particularly unusual about the shape of the stem and leaf plot in Figure 4.3? What might be an explanation for this?*

≝ *Aerofoil Effectiveness*

The stem and leaf plot for the aerofoil data in Section 3.15, showing fuel consumption in km/litre with the aerofoil fitted, is reproduced in Figure 4.4. Generally, we are interested in whether the fitting of the aerofoil has improved fuel consumption and what the average fuel consumption is with the aerofoil fitted.

> **Q4.** *What can you say about the pattern of variability in fuel consumption?*
> **Q5.** *What is a rough estimate of the average fuel consumption for this truck?*
> **Q6.** *From the data, what do you think might be the average fuel consumption on all trucks of this type?*

Frequency	Stem	Leaf
1	14	8
1	15	4
0	15	
2	16	33
3	16	669
3	17	123
8	17	56677889
8	18	00112444
15	18	555566677888999
16	19	0001112223333344
12	19	555566667799
5	20	01134
5	20	57789
1	21	0
3	21	588
1	22	0
1	22	7
1	23	0

Unit	0.1	0.01
Stem increment		0.05

Figure 4.4   Fuel consumption with aerofoil fitted.

4.2   Distributions

As you have seen in the previous three examples, repeated measurements of any quantity nearly always vary from each other. We can build up a picture of this variation as we collect data, as in Figure 4.5. This sort of picture, which is often drawn with dots rather than blocks (called a *dot plot*), shows us how values cluster at different points on the scale. It is rather like a stem and leaf plot turned 90°.

As you collect more data, the picture becomes more refined. The pattern formed is called a *distribution*. In Figure 4.6 the picture of the data is approximated by a smooth curve, which then represents the distribution of the data. This distribution curve is a *model* for the data.

Distributions can differ in the centre (or location), the amount of variation (or spread), or the shape of the distribution or any combination of these three. Changes in location, spread, and shape are depicted in Figure 4.7.

Differences in shape are often more fundamental than differences in location or spread, in that they reflect inherently different patterns of variation, whereas

Figure 4.5   Building a picture of data.

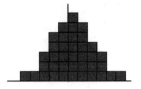 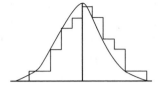 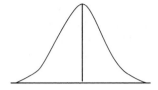

Figure 4.6   Distribution of the data.

Location   Spread   Shape

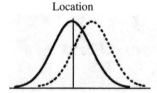 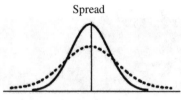 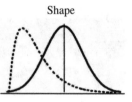

Figure 4.7   Distributions differing in location, spread, and shape.

a change in location or spread is just a slightly different manifestation of the same pattern. In the case of Foodstore, for example, the pattern of variation in savings might be essentially similar on different days of the week, with perhaps a difference in the average amount saved owing to the fact that weekend customers often tend to spend more, and hence save more, than on a weekday. However, if the store's policy were to change on how much to mark down or how many days before the sell-by date items are marked down, this could lead to a fundamentally different pattern of savings from customer to customer.

As is illustrated by the introductory examples to this chapter, a stem and leaf plot gives an indication of the properties of the distribution of the data. It can show whether the distribution is fairly symmetric or skewed; whether it has little variation or is very variable; or whether it has one, two, or more peaks. The stem and leaf plot is therefore a valuable aid in understanding the nature of the variation present in any process or situation.

Our aim in modelling is to find a distribution that provides us with a close approximation to data arising from the process under study. This usually involves determining quantities (parameters) that specify different features of the distribution, particularly location and spread. These parameters usually have to be estimated from the data using measures such as the mean and standard deviation, which are described in the following sections. We will return to a more detailed consideration of distribution shape in Section 4.8.

4.3   Arithmetic Mean

In addition to charts and diagrams, it is useful to summarise the data with a few well-chosen measures. The most important of these is an *average,* which measures the location of the data. The median, introduced in Chapter 3, is a useful form of average in some situations, but the most common form of average is the arithmetic mean, or simply the *mean*. This is calculated by adding up all the data values and dividing this sum by the number of values.

Consider the debt recovery data in Figure 4.1. There are 33 data values, from 69.0% in January 2000 to 79.0% in September 2002. The sum of these values is 2525.5, so

$$\text{Mean} = \frac{2525.5}{33} = 76.53\%$$

Q7. *Would you expect the mean percentage to increase or decrease if the January results were excluded?*

Q8. *Verify your answer by calculating the mean of the values, excluding the January figures.*

4.4 Standard Deviation

As well as defining a distribution by its location, we also need a measure to describe its spread. Consider again the debt recovery data in Figure 4.1. We can represent the data as a run chart with a horizontal line at the mean, as in Figure 4.8. Suppose we also draw a vertical line from each data point to the mean.

These deviations between the data values and the mean can be used to measure the spread or variation in the data. The greater the deviations, on average, the greater is the variation. The 33 deviations are given in Figure 4.9. Notice that some of the deviations are positive, while others are negative.

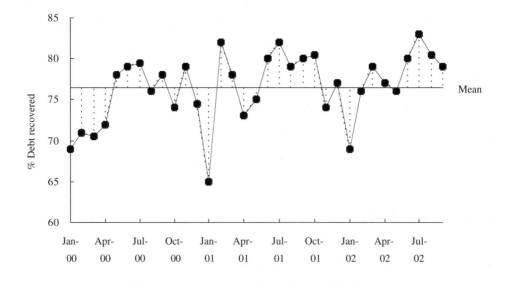

Figure 4.8   Debt recovery run chart showing deviations.

Month	2000	2001	2002	Month	2000	2001	2002
Jan	-7.53	-11.53	-7.53	Jul	+2.97	+5.47	+6.47
Feb	-5.53	+5.47	-0.53	Aug	-0.53	+2.47	+3.97
Mar	-6.03	+1.47	+2.47	Sep	+1.47	+3.47	+2.47
Apr	-4.53	-3.53	+0.47	Oct	-2.53	+3.97	
May	+1.47	-1.53	-0.53	Nov	+2.47	-2.53	
Jun	+2.47	+3.47	+3.47	Dec	-2.03	+0.47	

Figure 4.9   Deviations from the mean.

Q9.   What do you think the sum of these deviations will be? What is the reason for this?

The deviations tell us how far away each value is from the centre of the data. But if we simply average them to measure the spread of the data, we will always get zero. What we are interested in is the magnitude of the deviations, not their sign. One way of overcoming the problem with the sign is to square each deviation. Then we can take the mean of these squared deviations:

$$\text{Sum of squared deviations} = 581.97$$

$$\text{Mean of squared deviations} = \frac{581.97}{32} = 18.19$$

Finally, if we want our measure of spread to be in the same units as our data (% in this example), we take the square root of this number. The resulting measure of spread is called the *standard deviation*. For our example, we get

$$\text{Standard deviation} = \sqrt{18.19} = 4.26\%$$

✎ *Degrees of Freedom*

In calculating the mean of the squared deviations, we divided the sum by 32 rather than by the number of data values, 33. Why do we do this? Usually we are using a set of data to model the distribution of all data from a system or process. Ideally we would like to calculate the deviations using the mean of all the available data, not just of this small sample. However, the true mean is usually not available, so instead we must use an estimate of it from the data.

Because the deviations from the mean of a sample of data will always sum to zero, they are not all *independent*. For example, the sum of the first 32 deviations in Figure 4.9 is −2.47, and, since the sum of all 33 is zero, the last deviation must be +2.47. Hence, there are only 32 independent deviations. The condition that the deviations sum to zero means that we lose one *degree of freedom*. In general, to calculate the standard deviation from a sample of n values we divide the sum of the squared deviations by $n - 1$, the number of degrees of freedom, or independent deviations.

Later in the book we extend the idea to situations where there is more than one condition. For each condition, or parameter, that we have to estimate from the sample data, we lose a single degree of freedom.

✍ Excel Demonstration

To illustrate the process of calculating a standard deviation, we have provided on the CD-ROM a demonstration of the steps just described and a run chart illustration similar to that in Figure 4.8. The spreadsheet is called *SD.xls*, a copy of which is shown in Figure 4.10.

The data shown, from the debt recovery example, are entered into the first column of the spreadsheet under the heading *Value*. Only the first 20 values are visible in Figure 4.10. A run chart of the data is produced automatically as the values are entered. The steps in calculating the standard deviation are

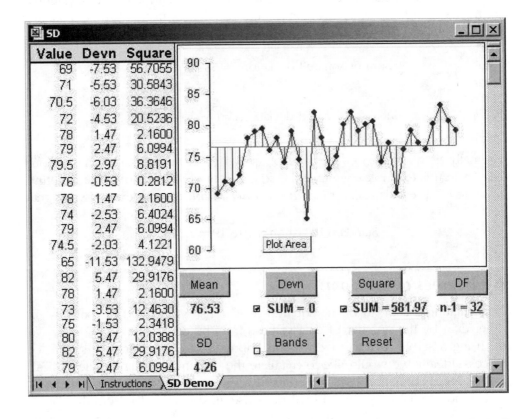

Figure 4.10   The standard deviation demonstration spreadsheet.

automated by the buttons under the graph. Clicking the *Mean* button calculates the mean value of the data (76.53) and draws the mean line on the graph. The deviations from the mean are then obtained and graphed by clicking the *Devn* button; they are also shown in the second column. Checking the small box below and to the left of the *Devn* button confirms that the sum of the deviations is zero. The squared deviations are calculated by clicking the *Square* button, and the sum of the squares is displayed by checking the small box below and to the left of this button. The square of each deviation appears in the third column. Finally, clicking the *DF* button confirms the degrees of freedom $(n-1)$ as 32, and clicking the *SD* button produces the value of the standard deviation (4.26).

Use of the *Bands* button will be discussed in Section 4.6 on interpreting the standard deviation. Clicking the *Reset* button will clear all the operations that have been carried out.

4.5 Calculating the Mean and Standard Deviation

As illustrated in the previous section, once the mean has been determined, we calculate the standard deviation in four steps:

1. Calculate the deviation of each value from the mean.
2. Square each deviation and calculate the sum of the squared deviations.
3. Divide by the number of degrees of freedom.
4. Take the square root.

This is tedious to do manually, but many electronic calculators have built-in functions for the mean and standard deviation. However, care has to be taken to ensure that the calculation of the standard deviation involves the correct number of degrees of freedom, because some calculators use the sample size, n, rather than the degrees of freedom, $n - 1$.

✎ *Using Excel*

Alternatively, we can use standard Excel functions to calculate the mean and standard deviation. The first step is to enter the data into an Excel spreadsheet. The mean and standard deviation are then obtained from the functions *AVERAGE* and *STDEV*.

For example, suppose that all the debt recovery data have been entered in cells A1 to A33. To obtain the mean, select a blank cell and enter the formula

= AVERAGE(A1:A33)

Likewise, in another blank cell, enter

= STDEV(A1:A33)

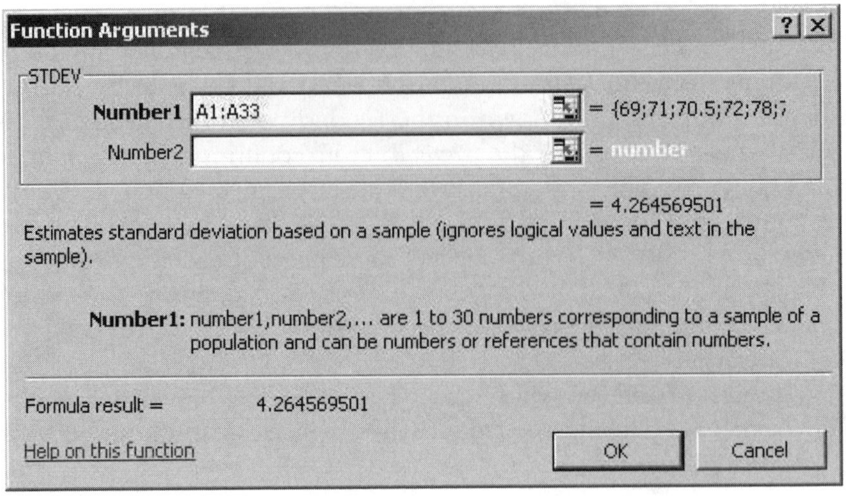

Figure 4.11   Using the STDEV function.

Rather than type in the function names, they can be selected from the *Insert... Function* menu. The dialogue box for *STDEV* is shown in Figure 4.11. The data range (A1:A33) is entered in the *Number1* box, and the formula result of 4.26 is the standard deviation, as before. The dialogue box for *AVERAGE* is virtually identical.

4.6   Interpreting the Standard Deviation

What does a standard deviation of 4.26% for the debt recovery data mean? If the distribution of the data is fairly symmetric, then we can say that:

- Approximately two-thirds of the data will lie within 1 standard deviation of the mean.
- Approximately 95% of the data will lie within 2 standard deviations of the mean.
- Nearly all the data will lie within 3 standard deviations of the mean.

For the debt recovery data, we would expect about two-thirds of the data to lie between 76.53 − 4.26% and 76.53 + 4.26%, that is, between 72.3% and 80.8%.

Q10. How many of the 33 data points fall within these limits?
Q11. What percentage of the total does this represent?
Q12. How does the actual percentage falling within 2 standard deviations compare with the expected proportion of about 95%?

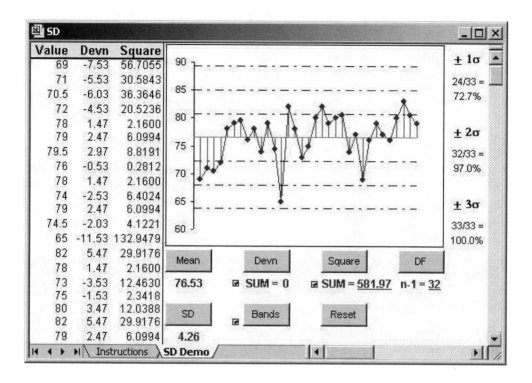

Figure 4.12   Demonstration of the 1, 2, and 3 standard deviation bands.

To check the accuracy of these results, we can again use the *SD.xls* demonstration spreadsheet. Clicking the *Bands* button draws in pairs of horizontal lines at 1, 2, and 3σ (i.e., standard deviations) each side of the mean. So the inner bands are at 76.5 ± 4.26, the middle bands are at 76.5 ± 8.53, and the outer bands are at 76.5 ± 12.79. Further, checking the small box to the left of the *Bands* button causes the percentage of values within each pair of bands to be determined, which is shown to the right of the graph.

This is illustrated in Figure 4.12, where it can be seen that about 73% of values actually lie within 1 standard deviation and 97% within 2 standard deviations and that no values lie outside 3 standard deviations.

✥ *Coefficient of Variation*

The standard deviation is a useful measure for comparing the variation in different data sets. However, it can be misleading to compare standard deviations in sets of data with different means. For example, suppose we are comparing the return on equities in two different sectors. In the communications sector the average return is 12% and the standard deviation is 4%, whereas the mean and standard deviation of the return on manufacturing stocks are 6% and 3%, respectively. The standard deviation of return from communications stocks is clearly greater, but do these stocks have greater risk? Relative to the average return, there is proportionately greater variability in manufacturing stocks, so arguably these present a greater risk. For this reason, it is common to express the standard deviation as a proportion (or percentage) of the mean. This gives us the *coefficient of variation* (CV).

For the debt recovery data, the coefficient of variation is

$$CV = \frac{4.26}{76.53} = 0.056 \qquad \text{i.e., } 5.6\%$$

Recall from Chapter 2 that a team was set up to work on the debt recovery problem. Data collected after the various improvement measures were put in place showed that the mean had increased to 81.6%, with a standard deviation of 3.16%.

> *Q13. What is the coefficient of variation after the improvement process?*
> *Q14. What effect has the improvement process had on the variability of the percentage debt recovered?*

4.7 Summary Statistics

As well as the mean and standard deviation, there are many other ways in which data can be described: other averages, such as the median discussed in Chapter 3, and other ways of measuring the spread of the data. For example, the range of the data and the interquartile range (also discussed in Chapter 3) are measures of spread, as is the mean absolute deviation (MAD).

Like the standard deviation, the MAD is calculated from the deviations. However, rather than square the deviations, we simply regard all the deviations as being positive. For example, in Figure 4.9 MAD is obtained by taking the mean of the 33 deviations, ignoring the minus signs. This gives a MAD of 3.42%. MAD can be calculated very easily using the Excel function *AVEDEV*, which works in exactly the same way as *AVERAGE* and *STDEV*.

> *Q15. For the debt recovery data, the mean absolute deviation is smaller than the standard deviation. Do you think this is just chance, or is there some reason this will usually be the case?*

Many of the more frequently used measures for describing data are provided as Excel functions, but to calculate them all individually is rather tedious and time consuming. Within the *STiBstat* add-in, described in Chapter 3, is a menu

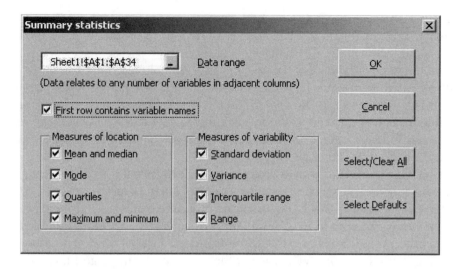

Figure 4.13   The *Summary Statistics* dialogue window.

item called *Summary Statistics*. This facilitates the calculation of the more commonly used descriptive measures by allowing you to select the ones you want and to calculate them all together. Figures 3.14 and 3.15 gave the *Summary Statistics* dialogue box and some summary statistics for the data in Figure 3.8. Figure 4.13 gives the dialogue box for the calculation of summary statistics for the debt recovery data.

Suppose the label "% Debt recovered" has been entered into cell A1 and the debt recovery data into cells A2:A34. In the *Summary Statistics* dialogue window in Figure 4.13, cells A1:A34 are entered as the *Data Range*, and *First row*

Summary Statistics

	% Debt recovered
No. of observations	33
Mean	76.53
Median	78
Mode	79
Lower quartile	74
Upper quartile	79.75
Maximum	83
Minimum	65
Standard deviation	4.265
Variance	18.187
Interquartile range	5.767
Range	18

Figure 4.14   Summary statistics for the debt collection data.

contains variable names is checked. In general, it is possible to determine summary statistics for more than one set of data at the same time, so the *Data Range* could contain more than one column. In this case Excel will always assume that the variables are located in different columns rather than in rows. Next we need to select which measures we want calculated by checking the appropriate boxes before finally clicking the *OK* button. If you want all the available measures calculated, simply click the *Select/Clear All* button. The results are written in a separate worksheet called *Output*, which will be automatically created if it does not already exist.

Using *Summary Statistics* for the debt collection data gives the output in Figure 4.14, some of which you will recognise, such as the mean of 76.53 and the standard deviation of 4.26. Many of the other measures need no explanation, such as the minimum, the maximum, and the range. The median, quartiles, and interquartile range have already been introduced in Chapter 3. The *Mode,* the most frequently occurring value, is another form of average, and the *Variance* is simply the square of the *Standard Deviation.*

> **Q16.** *From Figure 4.14, you can see that the mean is smaller than the median, which is itself smaller than the mode. What does this tell you about the distribution of the debt collection data?*

It should be noted that Excel has a similar built-in procedure for calculating various descriptive measures. It is a facility within the Excel *Data Analysis* tool-pack called *Descriptive Statistics*. You may need to load the *Data Analysis* add-in into the *Tools* menu if it is not initially available. The procedure for doing this is described in Section 8 in the *Introduction to Excel* on the CD-ROM. However, we prefer our own *Summary Statistics* add-in because some measures (such as the quartiles and the interquartile range) are not given in *Descriptive Statistics*, and also you can choose precisely what you want to have.

4.8 Distribution Shapes

As can be seen from the three examples at the beginning of this chapter, different shapes and patterns arise in the distributions of naturally occurring sets of data. Some situations tend to give rise to symmetrical distributions, whereas other distributions are *skewed* toward one end of the range. For example, the aerofoil data seem more or less symmetrical, whereas the Foodstore data are not. It is also often the case that distributions have a single peak (or *mode*), with progressively fewer values occurring further out from the centre, a sort of "bell-shaped" distribution. However, in contrast to this, the Foodstore data show evidence of two clusters of data with two separate peaks. Nevertheless, the data arising from many real-life processes are often more or less symmetrical and roughly bell shaped. The aerofoil data are one such example, but other examples of approximately symmetrical and bell-shaped data would be:

- The heights of males in Australia
- The weight of $2 coins
- The time taken by an experienced machinist to manufacture a standard component
- The scores from a large group of students taking a statistics test

However, there are many situations where the distributions will probably *not* be symmetrical and bell shaped, for instance, the distribution of:

- Personal incomes
- Waiting times for hospital treatment
- Time periods between accidents in a sawmill
- The time taken to drive from the centre of a large city to a particular suburb

Q17. For each of these last four examples, describe the shape of distribution you think will arise. In each case indicate why a symmetrical and bell-shaped distribution is unlikely.

White Rose Insurance Group

As part of its recruitment procedures for secretarial staff, the White Rose Insurance Group asks short-listed applicants to word process a standard document. Applicants are instructed to be most concerned about accuracy but nevertheless

Frequency	Stem	Leaf
1	7	0
4	7	5668
3	8	334
7	8	5567899
6	9	122224
6	9	566899
8	10	01123344
10	10	5566777899
6	11	001124
5	11	56689
2	12	01
2	12	59

Unit	1	0.1
Stem increment		0.5

Figure 4.15   Times to word process a test document.

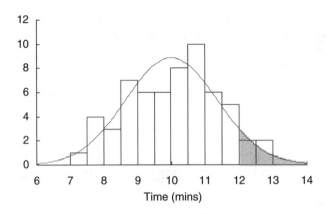

Figure 4.16   A model for the time to word process a test document.

aim to complete the task as quickly as possible. A maximum of 15 minutes is allowed. The times (in minutes) taken by a sample of 60 recent applicants for secretarial jobs are shown in Figure 4.15 as a stem and leaf plot and a histogram.

The stem and leaf plot in Figure 4.15 shows the shape of the data, but it is slightly clearer when presented in the form of a histogram. In modelling the distribution of a set of data, we are looking for a smooth curve that reflects the general characteristics of the data. If we collected many more observations, the detailed pattern of the data would almost certainly change slightly, so we want a curve that could reasonably reflect what a large sample of data might look like *in the long run*. In other words, we want a curve that represents the essential general characteristics of the data rather than the fine detail.

In Figure 4.16 we have superimposed a symmetric bell-shaped curve on the times in Figure 4.15. This curve provides a reasonable approximation to the bar chart; if we collected more data on times of other applicants, we might expect a bell-shaped curve to become an even better approximation of the bar chart. The curve, therefore, provides a good description, or model, of this situation.

From the stem and leaf plot in Figure 4.15, it seems that all applicants are able to complete the test document in less than the specified 15 minutes. It has been suggested, therefore, that it would be a more discriminating test if the time were reduced to 12 minutes. From the stem and leaf plot in Figure 4.15, it can be seen that 4 out of 60 applicants (i.e., 6.7%) took longer than 12 minutes, but would the model provide a more reliable estimate of this proportion? This is given by the area under the curve to the right of 12 minutes, as shown by the shaded area in Figure 4.16. However, to determine the model proportion, we must specify exactly the form of the distribution curve. A suitable model to use in this situation is the *normal* distribution, which is described in the next section.

4.9 The Normal Distribution

The most commonly used symmetric bell-shaped distribution curve is the *normal distribution*. The normal distribution is important and useful because:

- Many patterns of variation are naturally normal.
- The normal distribution is easy to work with mathematically and its properties are well known.
- In many applications conclusions drawn on the basis of the assumption of a normal distribution model are not seriously influenced by departures from the normal curve.
- Whenever a number of variable quantities is summed (or expressed as a mean), the result tends to be approximately normally distributed.

Although the equation that describes the normal distribution is complex, for our purposes the key point is that the normal curve is completely specified by the values of two quantities, or *parameters*. If we know the values of these parameters, then we can find the area under the normal curve between any two limits, which represents the proportion of observations that should fall in that interval. These two parameters are:

- The (arithmetic) mean, which will be denoted by the Greek letter μ (mu)
- The standard deviation, denoted by the Greek letter σ (sigma)

The mean and standard deviation of the times in Figure 4.15 are 10.0 min and 1.35 min, respectively. The curve superimposed on the histogram in Figure 4.16 is in fact a normal distribution with a mean of 10.0 and a standard deviation of 1.35. So if we can assume that the mean of *all* applicants' times will have a similar mean and standard deviation, a reasonable model for the time to word process the test document is a normal distribution with a mean of about 10 minutes and a standard deviation of about 1.35 minutes. But how do we calculate the proportion of applicants who take longer than 12 minutes?

✎ *Using Excel*

We can use the Excel function *NORMDIST* to calculate the required areas from a normal distribution. Figure 4.17 shows the *NORMDIST* dialogue box for calculating the proportion of values *less than* $x = 12$ for a normal distribution with $\mu = 10$ and $\sigma = 1.35$. The fourth quantity (*Cumulative*) must be specified as *true* (or any value other than 0) or else the function will give the height of the normal curve above the horizontal axis, which is not what we want.

Figure 4.17 gives the proportion of applicants who take less than 12 minutes, namely, about 93%. The remaining applicants, about 7%, will take more than 12 minutes. That is, we subtract the value in Figure 4.17 from 1 to give $1 - 0.931 = 0.069$, or about 7%.

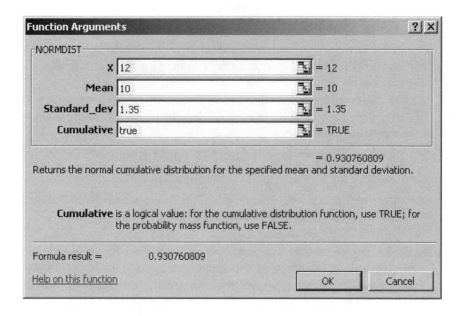

Figure 4.17   *NORMDIST* dialogue box.

> **Q18.** *What is your estimate of the proportion of applicants who would complete the test in less than 8 minutes?*

✍ *Using Normal Tables*

Traditionally statistics has used sets of tables to evaluate areas under the normal curve, rather than an Excel function. Normal distributions come with different means and standard deviations, but we can always transform them to an equivalent *standard normal distribution,* which has a mean of zero and a standard deviation of 1. The standard normal distribution has been extensively tabulated and is given in Appendix A.1. The table gives the proportion of items *greater than* any value z. Note that this is the complement of the probability given by *NORMDIST;* in other words, the *NORMDIST* value is 1 minus the value in Appendix A.1, and vice versa.

> **Q19.** *In a standard normal distribution, what proportion of values will exceed 1? What proportion will exceed 2? What proportion will exceed 3?*
>
> **Q20.** *Look back at the interpretation of the standard deviation in Section 4.6. Can you see the basis for the statements about the proportion of values falling within 1, 2, and 3 standard deviations of the mean?*

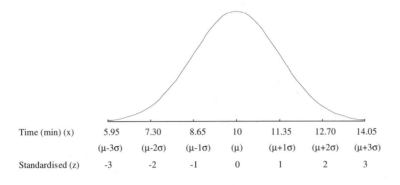

Time (min) (x)	5.95	7.30	8.65	10	11.35	12.70	14.05
	(μ-3σ)	(μ-2σ)	(μ-1σ)	(μ)	(μ+1σ)	(μ+2σ)	(μ+3σ)
Standardised (z)	-3	-2	-1	0	1	2	3

Figure 4.18   Standardising a normal distribution.

How do we proceed if the mean is not zero or the standard deviation is not 1?

In the White Rose Insurance Group data, we have a *non*standard normal distribution with a mean of 10 minutes and a standard deviation of 1.35 minutes. How can we use the table of the standard normal distribution to calculate the proportion of applicants with test times greater than 12 minutes?

Any normal distribution can be rescaled to the standard normal distribution, as illustrated in Figure 4.18. The procedure involves expressing any value, such as 12, as a difference from the mean in multiples of the standard deviation. This gives a *z-value* for the figure concerned. For example, if the mean is 10 and the standard deviation is 1.35, then a time of 12 minutes is 2 minutes, or 1.48 standard deviations, *above* the mean, in other words, a *z*-value of +1.48. In general, for any value *x*, its *z*-value is given by:

$$z = \frac{x - \mu}{\sigma}$$

where μ is the mean and σ is the standard deviation.

Thus, in White Rose, the proportion of test times over 12 minutes is the same as the proportion over 1.48 in the standard normal distribution. From Appendix A.1, this is 0.0694.

Formally, the calculation would be written as follows:

$$z = \frac{12-10}{1.35} = 1.48$$

Therefore, P(Test time > 12) = $P(z > 1.48)$ = 0.0694. That is, about 7% of applicants would take at least 12 minutes to complete the test, as before.

A Normal Distribution Model for the Aerofoil Data

From the shape of the stem and leaf plot in Figure 4.4, it appears that the fuel consumption data closely follow the pattern of a normal distribution. Furthermore, it is reasonable to assume that fuel consumption should vary symmetrically around some average figure, with higher and lower values becoming progressively less likely. So a normal distribution is a plausible model that fits the data.

The mean and standard deviation of the data in Figure 4.4 are approximately 1.9 and 0.15, respectively. Based on these values, we can set up a normal model for the aerofoil data. We can then use the model to answer questions such as the following:

- What proportion of trips has a fuel consumption between 1.7 and 2.2 km/litre?
- What are the quartiles of the distribution, and hence what is the interquartile range?
- What fuel consumption figure is exceeded on 90% of trips?

What Proportion of Trips Has a Fuel Consumption between 1.7 and 2.2 km/litre?

We first need to determine the proportions corresponding to $x = 1.7$ and $x = 2.2$, as shown in Figure 4.19.

For $x = 1.7$,

$$z = \frac{1.7 - 1.9}{0.15} = -1.33$$

$$P(x < 1.7) = P(z < -1.33) = 0.0918$$

Note that, since the distribution is symmetric, the proportion less than -1.33 is equal to the proportion greater than $+1.33$.

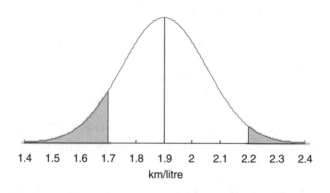

Figure 4.19   Probability that fuel consumption is less than 1.7 or greater than 2.2.

For $x = 2.2$,

$$z = \frac{2.2 - 1.9}{0.15} = 2.00$$

$$P(x > 2.2) = P(z > 2.0) = 0.0228$$

Adding these proportions ($0.0918 + 0.0228 = 0.1146$) gives the proportion below 1.7 and that above 2.2. The proportion of trips that has a fuel consumption between 1.7 and 2.2 is therefore $1 - 0.1146 = 0.8854$, or 88.5%.

What Are the Quartiles of the Distribution, and Hence What Is the Interquartile Range?

To determine the upper quartile, we must find a value of fuel consumption (x) that is exceeded 25% of the time. From Appendix A.1, first find the value of z that gives a proportion of 0.25. Searching in the body of the table, the nearest entry to 0.25 is 0.2514, corresponding to $z = 0.67$. The next entry is 0.2483, corresponding to $z = 0.68$, so a more accurate estimate of z is 0.675. The upper quartile is therefore 0.675 standard deviations above the mean, i.e.,

$$Q_3 = 1.9 + 0.675 \times 0.15 = 2.0 \text{ km/litre}$$

This is shown diagrammatically in Figure 4.20.
   Likewise, the lower quartile is 0.675 standard deviations below the mean, i.e.,

$$Q_1 = 1.9 - 0.675 \times 0.15 = 1.8 \text{ km/litre}$$

giving an interquartile range of $2.0 - 1.8 = 0.2$ km/litre.
   The quartiles of a distribution are examples of what are called *percentiles*. In particular, Q_1 is the 25th percentile and Q_3 is the 75th percentile. Percentiles for a normal distribution can be simply determined using an Excel function

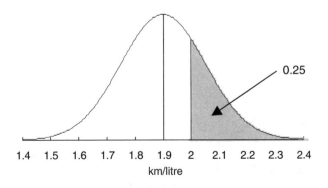

Figure 4.20   Fuel consumption that is exceeded 25% of the time.

Function Arguments	? X

NORMINV

Probability	.25	= 0.25
Mean	1.9	= 1.9
Standard_dev	0.15	= 0.15

= 1.798826571

Returns the inverse of the normal cumulative distribution for the specified mean and standard deviation.

Standard_dev is the standard deviation of the distribution, a positive number.

Formula result =   1.798826571

Help on this function   OK   Cancel

Figure 4.21   *NORMINV* dialogue box.

NORMINV. It gives the required value for a specified proportion (*probability*), mean, and standard deviation. The *NORMINV* dialogue box for finding the lower quartile is shown in Figure 4.21; the result is 1.8 km/litre, as before. Notice that the probability figure is again the proportion *below* the required value, and so the upper quartile corresponds to a probability figure of 0.75.

What Fuel Consumption Figure Is Exceeded on 90% of Trips?

This is another example of a percentile, this time a 10th percentile. This can be determined directly from Excel, with a figure of 0.1 entered as the *Probability* in Figure 4.21. The diagram is shown in Figure 4.22.

Alternatively, using the normal table, we require first the z-value corresponding to a proportion of 0.1. From Appendix A.1, the closest proportion in the

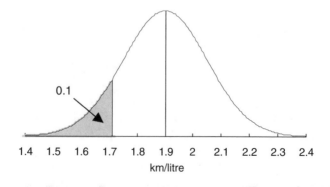

Figure 4.22   Fuel consumption that is exceeded 90% of the time.

table is 0.1003, corresponding to $z = 1.28$, but this is the value that is *exceeded* 10% of the time. The value we require is therefore $z = -1.28$, i.e., 1.28 standard deviations *below* the mean. Hence, the fuel consumption figure exceeded on 90% of trips is $1.9 - 1.28 \times 0.15 = 1.71$ km/litre.

If you determine these quantities from the data in Figure 4.4, you will find that:

- Seventy-seven of the 86 values are between 1.7 and 2.2, which is 89.5%.
- The lower quartile (the 21.75th value) is 1.81 and the upper quartile (the 65.25th value) is about 1.96. This gives an interquartile range of about 0.15.
- Eight values (9.3%) are 1.71 or less, and nine values (10.5%) are 1.72 or less. Thus the 10% point is about 1.715.

There is, therefore, fairly good agreement between the data and the results based on a normal model, which perhaps justifies our use of a normal distribution. The results from the normal model should be more representative of the overall situation, provided of course that our model is the correct one.

> *Q21. The manufacturers of the aerofoil have brought out a new design that is claimed to increase average fuel performance (more km/l) by 10%. Assuming their claim is correct, how would this alter your answers to the three questions?*

✍ *Percentage Points of the Normal Distribution*

Consider again the times taken to complete the word processing test by secretarial applicants to White Rose Insurance Group. Most applicants take between 8 and 12 minutes. Assuming processing times follow a normal distribution with mean 10 minutes and standard deviation of 1.35 minutes, it can be shown that the exact proportion of applicants having times within 2 minutes of the mean is $1 - 2 \times 0.069 = 0.862$, i.e., about 86%. However, suppose the recruitment manager wishes to determine the value of x such that exactly 90% of applicants are within x minutes of the mean. In other words, 5% of people tested will take less than $10 - x$ minutes and 5% will take longer than $10 + x$ minutes.

As in the aerofoil example, we are looking for two percentiles of the distribution, namely, the 5th percentile and the 95th percentile. Due to the symmetry of the normal distribution, both percentiles correspond to a *P*-value of 0.05 in Appendix A.1. Examining the entries in Appendix A.1 shows us that a *z*-value of 1.64 leads to $P = 0.0505$ and $z = 1.65$ gives $P = 0.0495$. Thus a *P*-value of 0.05 corresponds to $z = 1.645$. In other words, 90% of applicants will have test times within 1.645 standard deviations of the mean, that is, within $1.645 \times 1.35 = 2.2$ minutes of the mean, or between $10 - 2.2 = 7.8$ minutes and $10 + 2.2 = 12.2$ minutes. The value $z = 1.645$ is called the 90 percent point of the standard normal distribution. More generally, we may wish to know the value of z that

has a specific percentage (*P*) of the standard normal distribution falling between
−*z* and +*z*. A selection of these *percentage points*, and a diagrammatic representation, is given in Appendix A.2.

4.10   Case Study: Julia's Diner

Julia Roberts, the owner of a local diner, is concerned about the time it takes
to serve the food, once the order has been taken, during the busy lunchtime
period between 12 noon and 2 p.m. To pacify customers who wait a long time,
she wonders whether it would be a good idea to give a $20 voucher to those
customers who have waited for at least 30 minutes. She has an average of 220
lunchtime customers each week. If she goes ahead with her idea, she needs to
be sure the scheme will not prove too expensive. Consequently, she collects
data over two particular days on the time taken to serve a total of 97 lunchtime
orders. The data are presented in the stem and leaf plot in Figure 4.23.

> *Q22. What proportion of the 97 customers would have received vouchers?*
> *Q23. What would be the expected weekly cost of the vouchers to Julia?*

Frequency	Stem	Leaf
1	1	0
4	1	2233
6	1	445555
6	1	666777
8	1	88899999
12	2	000000111111
13	2	2222222233333
16	2	4444444455555555
14	2	66666667777777
9	2	888888999
5	3	00001
3	3	233

Unit	10	1
Stem increment	2	

Figure 4.23   Stem and leaf plot of waiting times.

Your answers to these questions are determined from the limited sample of data in Figure 4.23 and so may not really reflect what is actually happening. If we use these data to develop an appropriate *model* of the service times, then it may be possible to calculate these quantities more accurately. On examining the shape of the stem and leaf plot in Figure 4.23, we find clear evidence of a bell-shaped pattern. The distribution is almost symmetrical, although there is some evidence of a small amount of skewness in the results, with the peak (mode) being slightly above centre. This is described as negative skewness.

> *Q24. What might be a possible explanation for the slight negative skewness in Figure 4.23?*

Despite the suggestion of skewness in the data, a normal distribution should be a reasonable model for waiting times in Julia's diner. We therefore need to determine the mean and standard deviation of waiting time so as to specify the particular normal distribution required. Use of the appropriate Excel functions (or the *Summary statistics* add-in) gives values of 22.7 and 5.13 minutes for the mean and standard deviation, respectively. To demonstrate how well the normal distribution fits the data, the normal curve with these parameters is superimposed on a histogram of the data in Figure 4.24. So to estimate the proportion of waiting times that are greater than 30 minutes (the discount qualifying time) we can use the model given by this curve. This proportion is given by the area under the curve to the right of 30 minutes, as shown by the shaded area in Figure 4.24. The calculation proceeds as follows.

For $x = 30$,

$$z = \frac{30 - 22.7}{5.13} = 1.42$$

$$P(x > 30) = P(z > 1.42) = 0.078$$

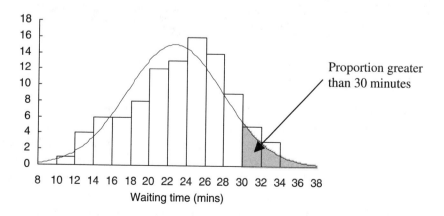

Figure 4.24   Distribution of service times.

Therefore, the probability that waiting time exceeds 30 minutes is 0.078; that is, about 8% of customers should have to wait for at least 30 minutes for their meal to be served. This value corresponds very well with the data, where 8 out of 97 customers waited 30 minutes or longer, i.e., a proportion of 8/97 = 0.082. In general, however, if the model we have used is an appropriate one, we would expect the model proportion to be a more accurate figure because it "smoothes out" the irregularities in the data and also overcomes the issue that time is a continuous quantity whereas the times recorded in the data are to the nearest minute. When estimating the proportion from the data, we also have a problem of how to deal with the four values of exactly 30 minutes that were recorded: Do they count as being over 30 minutes or not?

Due to the potentially high cost of the vouchers, Julia decides that a more effective solution to the problem may be to improve service at the diner. She has two strategies in mind. One is to employ an extra lunchtime waiter, which should reduce the average service time to 15 minutes. The extra waiter will cost about $200 a week. Her other idea is to reorganise the kitchen operations at a one-off cost of $1000, which should reduce the variability in service time by about 40%. Which strategy should she pursue if she continues the policy of giving $20 vouchers for service that takes at least 30 minutes?

An extra waiter reduces the average time to 15 minutes with (we assume) the same standard deviation. For the proportion greater than 30 minutes, $z = (30 - 15)/5 = 3$ and $P(z > 3) = 0.0013$. The expected weekly cost of vouchers is now $0.0013 \times 220 \times 20 = \$5.72$, a saving of approximately $346 a week. At a cost of $200 a week, the extra waiter looks like a good idea.

The reorganised kitchen reduces the standard deviation to $0.6 \times 5 = 3$ minutes, with (we assume) the same average as before. For the proportion greater than 30 minutes, $z = (30 - 23)/3 = 2.33$ and $P(z > 2.33) = 0.01$. The expected weekly cost of vouchers is now $0.01 \times 220 \times 20 = \$44$, a saving of approximately $308 a week. At a cost of $1000, the kitchen reorganisation would pay for itself in under a month and also looks like a good idea.

> **Q25.** *Which of the two options would you advise, and why?*

4.11 Binary Data

Attribute data, defined in Section 3.1, classifies each item into one of a number of mutually exclusive categories. In particular, attribute data with only *two* categories frequently arise in practice. For example, a coin is a head or a tail, a person is male or female, and an inspected product passes or fails a test. Such situations, where items can be thought of as possessing a certain attribute (such as a person being male or a coin falling heads), or not, give rise to *binary* data. In the remaining sections of this chapter we will examine some of the distributional properties of binary data. The analysis of attribute data, in general, will be considered in Chapter 5.

It is usual to summarise the results of a set of binary data in terms of a measure of the frequency of occurrence of one of the categories. This measure will generally be either a *proportion* or a *count*. For example:

- The proportion of defective items in a batch, the proportion of red beads in the paddle, or the proportion of a day's production of computer chips that does not meet specifications.
- The number of computer breakdowns in a week, the number of defective welds in 100 m of pipeline, or the number of accidents in a year. All these are *counts* rather than proportions.

So when do we use one rather than the other? In particular, it seems we could equally well express proportions as counts. For instance, to work out the *proportion* of red beads, we clearly need first to have a *count* of how many were in a given sample of data. In practice, we tend to use a proportion when the attribute in question occurs so many times within a defined total number of observations. With the beads experiment, knowing that the paddle contains a total of 50 beads, the occurrence of (say) 12 red beads is more informative if expressed as a proportion of 12/50 = 0.24, or 24%.

However, there are some situations where the total number of possible occurrences is not meaningful or, more specifically, where nonoccurrence is not observable. Such situations often arise when the basis of observation is for a specified period of time rather than a given total number. For example, if the attribute (phenomenon) we are recording is accidents in a factory, then how can we observe a "nonaccident"? The number of potential accident situations is extremely large (infinite?), so we cannot reduce the number actually observed to a meaningful proportion. The best we can do is to count the number of accidents occurring in a specified time period, such as a week or a month.

To make matters even more complicated, in some situations proportions are used, but not arising from attribute data. In particular, we might be dealing with two related interval quantities measured in the same terms where it is useful to look at one in relation to the other. For example, it is often useful to look at the monetary value of stock in relation to total turnover. This gives a ratio (or proportion), but not one that arises from counting anything. It is simply the ratio of two interval quantities and is therefore itself an interval quantity.

4.12 Working with Proportions

In previous chapters, the importance of understanding variation has been stressed repeatedly, and this continues to be true when working with proportions. Just as the characteristics of a sample of interval data will vary from sample to sample, so will the proportion of some attribute, such as red beads or defective items. For example, when a coin is tossed five times, the proportion of heads obtained could be 0, 0.2, 0.4, 0.6, 0.8, or 1, corresponding to anything from 0 to 5 heads. Not all of these outcomes are equally likely, but they are all possible.

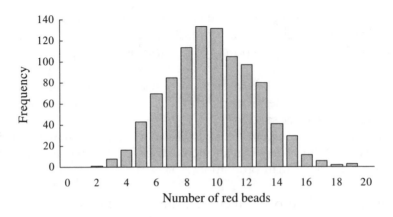

Figure 4.25   Results from 39 beads experiments.

If a coin is tossed fairly, we expect any sample to produce close to 50% heads and 50% tails, but there is likely to be variation around this figure from one sample to the next.

Consider the beads experiment, which we have conducted many times with New Zealand and British managers and with university students. A histogram of the results from 39 experiments, involving 975 individual results, is given in Figure 4.25. The proportion of red beads in the paddle could be anything from 0 to 1, but the evidence of Figure 4.25 seems to indicate that the proportion is unlikely to be greater than $20/50 = 0.4$. From the data in Figure 4.25, the mean number of red beads per batch is 9.73, with a standard deviation of 2.90. So we expect the proportion of red beads to be about $9.73/50 = 0.195$ (i.e., about 20%), with a standard deviation of $2.9/50 = 0.058$ (i.e., about 6%).

In general, we shall let p denote the *expected* proportion of items in a particular category, for example, red beads in the paddle or heads obtained when a coin is tossed a number of times. Usually p must be estimated from a set of data such as in Figure 4.25, although occasionally the value of p can be assumed from the characteristics of the situation involved. For example, it is reasonable to assume that a fair coin can be expected to show heads 50% of the time.

For simplicity let us assume we are concerned with the proportion of defective items in a batch (or sample) of a given size n (in the red beads experiment, $n = 50$). If the probability of getting a defective item remains constant and if the outcome of one item is unaffected by (or independent of) the outcomes of any previous items, then the probability of getting a given number of defectives in the batch is given by the *binomial distribution*. This distribution is covered in detail in most books on business statistics, but we shall only be concerned with two important properties of the distribution:

1. How the standard deviation of the proportion is influenced by p and n
2. The shape of the distribution for large sample sizes

✎ *Standard Deviation of a Proportion*

We know from the theory of the binomial distribution that if the expected proportion defective is p, then the standard deviation of the proportion defective in a random sample of n items is given by

$$\text{Standard deviation} = \sqrt{\frac{p(1-p)}{n}}$$

(The standard deviation of a proportion is often called the *standard error* of a proportion.)

This means that we can infer the value of the standard deviation from an estimate of the proportion p in question. For instance, if we know that the average number of red beads is about 10, this implies a proportion of about $p = 10/50 = 0.2$. We then get an estimate of the standard deviation of $\sqrt{0.2 \times (1 - 0.2)/50} = 0.0566$, which is in close agreement with the value 0.058 calculated from the data in Figure 4.25.

Figure 4.26 gives a plot of the standard deviation against p for the simplest case, where $n = 1$. The shape of the relationship in Figure 4.26 is the same for $n > 1$, where only the vertical scale changes.

> **Q26.** *Why do you think the standard deviation increases as p increases from 0, reaching a maximum at $p = 0.5$, and then decreases until it is equal to 0 at $p = 1$? Is this intuitively reasonable?*

✎ *Normal Approximation for Proportions*

From the shape of the histogram in Figure 4.25, it seems that the distribution of the number of red beads in the paddle follows quite closely the pattern of

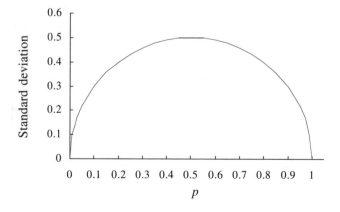

Figure 4.26 Standard deviation of a proportion.

the normal distribution. This would still be true if the data were expressed as proportions; for example, 10 red beads becomes a proportion of 10/50 = 0.2. Working with proportions simply changes the horizontal scale by a factor of 50, the total number of beads in the paddle. So, we can say that the proportion of red beads is approximately normally distributed with a mean of about 0.2 and a standard deviation of about 0.06. The normal distribution gives a good description, or *model*, for the outcome of the beads experiment. We could therefore use this model to estimate the likelihood of getting, say, more than 15 red beads, that is, a proportion greater than 0.3.

But is it just coincidence that the normal distribution gives a good approximation to the beads data, or is there a more general process at work here? It can be shown that the normal distribution gives a good approximation to any binomial distribution provided the sample size n is large. As you can see from the beads data, it is very good with $n = 50$. There are some other rules of thumb that should be considered before using this normal approximation, which are related to p, the proportion of defectives. The normal approximation is very good, even for small n, when p is close to 0.5. You should exercise caution, however, when p is very small or very large, for then the normal does not provide a good approximation.

To further illustrate the normal approximation, let us take samples from a situation where the probability of getting a defective item is $p = 0.5$. As an analogy, we can think of an experiment where we toss a coin a number of times and we are interested in the proportion of heads. The histograms for samples of sizes $n = 1, 2, 4,$ and 8 are given in Figure 4.27. You can see that, even with only a sample of size $n = 4$, the "bell-shaped" pattern of a normal

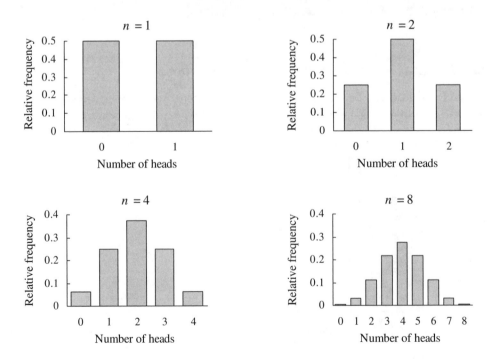

Figure 4.27   Bar charts for samples of size 1, 2, 4, and 8.

distribution is beginning to emerge. With a sample of size $n = 8$, the approximation to the normal distribution has already become quite good.

4.13 Working with Counts

Recall from Section 4.11 that count data usually arise when we are interested in the number of occurrences of some event during a specified period of time, such as the number of:

- Motorists caught by a speed camera
- Passenger complaints received by British Airways
- People entering the accident and emergency department at Waikato hospital
- Telephone calls arriving at the switchboard at Loughborough University
- Cars arriving at the toll booths at Sydney Harbour Bridge

In all these examples, the count in question will relate to some appropriate period of time.

Assuming that conditions stay the same over the period in question, in all of these situations the probability of a certain number of occurrences is given by the *Poisson distribution*. Details of this distribution can be found in most statistics books, but, as with the binomial distribution, we shall only be concerned with:

1. The relationship between its mean and standard deviation
2. A normal approximation for this distribution

✥ *Standard Deviation of a Count*

Suppose we are concerned with a count of some quantity, such as the number of insurance claims per year or defects per day or accidents per quarter. From the theory of the Poisson distribution, the standard deviation (or standard error) of this count is given by $\sqrt{\mu}$, where μ is the average rate of occurrence. For example, if the daily average number of defects is 25, then the standard deviation of the daily number of defects is 5. Figure 4.28 gives a plot of the standard

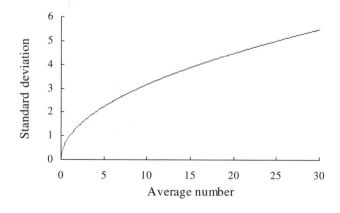

Figure 4.28   Standard deviation of a count.

deviation against the average number. It can be seen that as the average number increases, so does the variability as measured by the standard deviation.

Is this relationship between the average rate and the standard deviation intuitively reasonable? Consider the occurrence of accidents in the following two situations. In a small workshop, the average number of accidents is likely to be very low, say, about 1 per month, while in a large timber company the average might be 30 per month.

> **Q27.** *How many accidents each month might we typically get in these two situations?*
>
> **Q28.** *Would you conclude that when the accident rate is low we get little variability in our results, while a higher accident rate leads to greater variability?*

For a second example, consider the abundance of oysters in the seas around New Zealand. Some areas will have no oysters, that is, zero on average, with no variability. In other areas oysters will be rare, with very low numbers on average and, hence, little variability in the numbers found. Around Bluff, oysters are found in huge quantities, and the numbers found will vary enormously from day to day.

✍ *Normal Approximation for Counts*

Suppose we have collected data over a 6-month period (182 days) on the number of claims per day from an insurance company, giving the histogram in Figure 4.29. As with proportion data, this is similar in shape to a normal

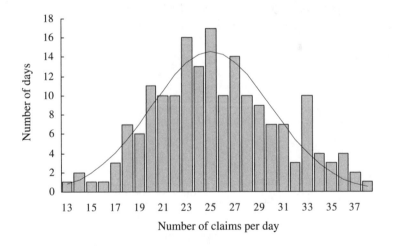

Figure 4.29   Normal curve on daily number of claims.

distribution, as can be seen from the superimposed normal curve. The fit here seems less good than in Figure 4.25, but it must be remembered that there were 975 individual results in Figure 4.25, whereas here there are only 182. If more data were available, it is likely that some of the "irregularities" in Figure 4.29 would disappear and the normal distribution would give a better model for the data.

From the data, the average number of claims is 25.5, with a standard deviation of 5.3, which conforms well to the Poisson distribution model (the standard deviation is approximately equal to the square root of the mean). However, as with the earlier beads data, it is unlikely that information about the standard deviation would be available in practice. But we can estimate it to be about 5 based on an assumption that the mean is about 25. Using the normal model, we can now estimate the probability of (say) more than 30 claims per day from the area under the normal curve above 30.5. This gives

$$z = \frac{30.5 - 25}{5} = 1.1$$

$$P(\text{claims} > 30) = P(z > 1.1) = 0.136$$

Thus, there should be more than 30 claims on about 14% of days. There were actually 34 days when the number of claims exceeded 30, a proportion of 18.7%. In this case, the model underestimates the actual proportion, as can be seen from the right-hand tail of Figure 4.29, where all but one of the values above 30 had a frequency greater than that predicted by the normal model. Had more data been available, it is likely that the actual proportion of days with more than 30 claims would be closer to the predicted value.

Q29. *Why do you think we used the area to the right of the value 30.5 when estimating the probability of more than 30 claims? If you were finding the proportion of days on which there are less than 20 claims, what normal distribution area would you use?*

Q30. *There were 21 days out of 182 (i.e., 11.5%) with fewer than 20 claims. From Figure 4.29, would you expect the normal model to overpredict or underpredict this proportion? Why?*

Q31. *Check that you are right by estimating the proportion of days with fewer than 20 claims based on a normal model with a mean of 25.*

In general, a normal distribution should provide a good model for count data as long as the average rate of occurrence is reasonably large. In most cases an average greater than 10 should be sufficient.

These relationships regarding the standard deviation of a proportion or a count will be exploited in Chapter 7, the section on margins of error for proportions, and in Chapter 12, the section on control charts for attribute data.

4.14 Case Study: Inter-Island Airways

Inter-Island Airways is a small airline that operates scheduled flights between various islands in the eastern Caribbean. Seats on all flights are standard economy class, and the terms of sale of most tickets allow passengers to cancel a booking at any time for a small administrative charge of $25. This has led many flights to operate at less than full capacity due to last minute cancellations and consequent loss of revenue.

Following a poor profit performance in the previous quarter, mainly because of subcapacity operation on many flights, the operations director decides to analyse a particularly problematic route to determine the proportion of bookings cancelled within 48 hours of departure, because it is very unlikely that a seat can be resold following such late cancellation. The route in question uses a Boeing 737 aircraft with a capacity of 150 passengers. The stem and leaf plot in Figure 4.30 gives the numbers of late cancellations from a sample of 54 recent flights that were "fully booked." It can be seen that the number of cancellations per flight ranges from 13 to 34.

The data involved here are binary, because a booking is either cancelled or not. Further, we have proportional data because the number of cancelled passengers (on a fully booked flight) can be expressed as a proportion of the aircraft capacity of 150 passengers. From the data in Figure 4.30, the average number

Frequency	Stem	Leaf
1	1	3
2	1	45
2	1	67
1	1	9
4	2	00011
10	2	2222333333
14	2	44444445555555
11	2	66666677777
4	2	8889
2	3	01
1	3	2
2	3	4

Unit	10	1
Stem increment	2	

Figure 4.30   Number of late cancellations per flight.

of late cancellations is 24 per flight, which corresponds to a proportion of 24/150 = 0.16.

A critical performance target for the airline is that for any route to be profitable it should have a load factor of at least 80%. Applying this figure to the route concerned, we should aim for no more than 20% cancellations or, equivalently, at most 30 empty seats. From the normal model, we can estimate the proportion of flights on this route that achieve an 80% load factor as follows. Since, the standard deviation of the proportion is

$$\sqrt{\frac{0.16 \times 0.84}{150}} = 0.03$$

then

$$P(\text{Load factor} > 0.8) = P(\text{Proportion seats cancelled} < 0.2)$$

$$= P\left(z < \frac{0.2 - 0.16}{0.03}\right) = P(z < 1.33) = 0.9082$$

Hence, we estimate that about 91% of all fully booked flights (on this particular route) will have a load factor greater than 80%. From the data in Figure 4.30, there are three flights with more than 30 empty seats, which means that 51/54 = 0.944 (i.e., 94%) of flights have a load factor greater than 80%.

With our model, we can now examine the impact of any changes that might be considered. For example, it has been suggested that the pricing policy could be changed whereby the fare is reduced by $20, but there is a significantly increased charge of $100 when passengers cancel their booking within 48 hours of the flight date. With the higher cancellation charge, it has been estimated that there will be a reduction of 40% in the number of late cancellations, but to offset the fare reduction, the load factor must increase to at least 85%.

> **Q32.** *With this price change, what is the chance of achieving the desired load factor of 85%? Does the pricing change look as if it likely to be profitable?*

4.15 Chapter Summary

This chapter has been concerned with identifying the key features of a set of data and developing an appropriate model of the data that reflects the inherent characteristics and shape of the distribution concerned. Specifically:

- How data distributions arise, and the main patterns and features that occur
- How location and spread can be measured by the arithmetic mean and standard deviation

- Interpreting the standard deviation
- The widespread occurrence of bell-shaped distributions, which are often symmetrical or nearly so
- The use of a normal distribution to model symmetrical, bell-shaped distributions
- How the normal distribution can be used to estimate the proportion of values that can be expected to occur in particular ranges within the distribution, and how these proportions can be calculated using either normal tables or Excel
- Using a normal distribution model to answer "what-if" questions regarding the likely consequences of proposed changes
- The nature of binary data and how the data can be summarised in terms of either a proportion or a count of how frequently a characteristic occurs in a sample of attribute data
- How the standard deviation can be determined from either the proportion of items with some characteristic or the average number of occurrences of some event
- The fact that proportions and counts tend to give rise to normal distributions under certain conditions, and how this can be used to determine probabilities in binary situations

4.16 Exercises

1. An automatic teller machine (ATM) situated on Victoria Street dispenses cash to bank customers. The amount of money dispensed each day is approximately normally distributed with a mean of $8500 and a standard deviation of $1500.
   a. If $11,000 is put into the ATM at the beginning of each day, on what proportion of days will the ATM run out of money?
   b. What percentage of days will the amount dispensed be between $7,000 and $10,000?
   c. What amount of money should be placed in the ATM at the beginning of each day so that the ATM runs out of money on no more than 1% of days?

2. A large company has an annual performance appraisal system to assess its employees. Each employee is assessed on a number of different criteria related to his or her performance. On each criterion the employee is given a score on a scale from 0 to 9, where 0 means very unsatisfactory and 9 means outstanding. These individual scores are then averaged to produce an overall rating for the employee. A rating of less than 3 is regarded as "underachieving," while a score greater than 8 is regarded as "exceptional." Underachievers and exceptional employees are the subject of special attention by senior management. In an attempt to ensure consistency, fairness, and comparability throughout the company, each departmental head has to ensure that the ratings of about two-thirds of the employees in his or her department are between 5 and 7. The managing director assesses departmental heads.

a. Assuming that the distribution of performance ratings can be approximated by a normal distribution with mean 6 and standard deviation 1, what proportion of employees are expected to underachieve? What proportion will be exceptional?
b. Comment on the performance appraisal system used by this company. What are its good and bad points, if any? Can you think of a better way of assessing performance?

3. A company manufactures a type of PVC pipe that has a minimum tensile strength specification of 52 MPa. There has been a change of supplier of raw materials, and the production manager wants to check the tensile strength of the pipes manufactured with the new materials. A sample of 50 pipes was tested and gave a sample mean of 65 MPa and a sample standard deviation of 4.8 MPa.
a. What proportion of the pipes made using the new materials has a tensile strength below the specification limit? State any assumptions you make.
b. With the previous supplier of the raw material the proportion of pipes with a tensile strength below 52 MPa was about 2%. Can you be confident that the change of supplier has led to an improvement?

4. Hill's Weaving Company is looking at two suppliers of yarn to be woven into men's suits. The elasticity of the yarn is an important attribute, because the yarn tends to snap if the elasticity is low. The elasticity of 40 samples of yarn has been measured from both suppliers. Supplier A's yarn has an elasticity with a mean of 28.3 and a standard deviation of 3.5, whereas supplier B's has a mean of 33.0 and a standard deviation of 5.5. If the elasticity is below 12, the yarn is unacceptable, because it will snap in use.
a. Which supplier would you choose?
b. Explain the reasoning for your choice.

5. A bank is looking into the amount of cash in new notes it should keep available for customer cash withdrawals. The bank always tries to pay out withdrawals in new notes. A recent study has shown that the amount of cash withdrawn by customers on any day is normally distributed with mean $14,000 and standard deviation $2,000.
a. Suppose the bank starts each day with $18,000 in new notes. On what proportion of days will old notes have to be paid to customers?
b. The bank aims to ensure that used notes are issued on only 1% of days. What amount of new notes is required at the start of each day to achieve this target?
c. The bank is looking at reducing the amount of new notes it holds each day. It is examining the following two strategies:
   i. To reduce the amount of withdrawals, the bank can use various initiatives to encourage its customers to use EFTPOS (paying for goods and services electronically at the point of sale). They estimate this will reduce mean withdrawals to $13,000 but leave the standard deviation unchanged.
   ii. Encourage customers to make withdrawals less often by charging higher bank fees for each withdrawal. They estimate this will leave the mean unchanged at $14,000 but reduce the standard deviation to $1,500.

If the bank wants to ensure that used notes are issued on only 1% of days, what strategy should it adopt?

6. A machine shop manager wishes to study the time it takes an assembler to complete a small subassembly. Measurements, in minutes, are made at consecutive half-hour intervals.

   a. Explain why a run chart might be an appropriate way of displaying these data.

   b. What graphical method would you use if you were interested in detecting skewness in these data?

   c. The mean time taken by the assembler is 12 minutes with a standard deviation of 2 minutes. Assuming the data follow a normal distribution, estimate how often the time taken would exceed 15 minutes?

7. Brian Hearse, the manager of Pine-Furniture in Hamilton, is looking into the viability of opening on Sundays. His daily sales on weekdays average $120,000, with a standard deviation of $8000. Brian decides that if he can be almost certain that his store will gross at least $50,000 on a Sunday, then it will be worth opening. However, he knows that a similar store in Wellington with average sales of $167,000 on weekdays only averages $98,000 on Sundays.

   a. What do you think will be a reasonable estimate of average Sunday sales in Hamilton, and why?

   b. Based on your estimate in (a), should Brian open the Hamilton store on Sundays? (Assume a standard deviation of $8,000 for Sunday sales and that sales are approximately normally distributed.)

   c. Brian is considering using an advertising campaign to boost the Sunday opening. Making the same assumptions, what mean sales should this advertising campaign achieve so that Brian can be 95% certain that sales will exceed $60,000?

8. The manager of a medical centre schedules appointments every 10 minutes for patients wishing to see the doctor. The time patients actually spend with the doctor is normally distributed with a mean of 8 minutes and a standard deviation of 1.7 minutes.

   a. What is the probability that a patient spends longer than 10 minutes with the doctor?

   b. Assuming that all patients arrive on time for their appointment, at what interval should appointments be scheduled if the manager wants to be 99% certain that a patient will not have to wait?

   c. Using the appointment interval in (b), how many patients would be seen by the doctor in the course of a 2-hour surgery? What would be the average doctor idle time? Do you think that this would be a sensible appointment system to use in practice? Explain the reasons for your answer.

9. The time customers have to wait to have a tyre replaced in a tyre service centre is normally distributed with a mean of 20 minutes and a standard deviation of 6 minutes.

   a. What is the probability that a randomly selected individual will have to wait at least 15 minutes to have a tyre replaced?

   b. In an effort to improve customer relations, all customers who have to wait longer than 30 minutes are given a voucher offering a 20% discount on their next tyre replacement. What proportion of customers will receive vouchers?

c. What is the average annual cost to the company of the voucher scheme if there are 2500 customers a year and the average amount they spend is $80 per customer? You should assume that 75% of the vouchers issued would be used against future purchases.

d. In order to reduce the cost of the voucher scheme, it is proposed to institute an incentive scheme for the tyre fitters (at a cost of $1000 per year). This should have the effect of reducing average waiting time to 15 minutes, but with the same standard deviation as before. Would this be a cost effective strategy?

10. Suppose that the waiting time for a particular service is normally distributed with a mean of 22 minutes and a standard deviation of 8 minutes.

a. What is the probability that a randomly selected individual will have to wait at least 30 minutes?

b. Following various customer complaints about excessive waiting time, management has issued a directive that no customer should have to wait longer than 30 minutes. To adhere to the directive, it has been decided that either the mean or the standard deviation should be reduced. Two particular strategies (A and B) have been identified that would result in the following reductions:

Strategy A. The mean is reduced to 15 minutes, and the standard deviation remains at 8 minutes.

Strategy B. The mean remains at 22 minutes, and the standard deviation is reduced to 5 minutes.

Which of these strategies would be more effective toward meeting the directive?

11. You have just opened a restaurant in Derby and you are trying to estimate the amount of ground coffee you should purchase each week. Weekly data show that coffee consumption is approximately normal with mean 4.4 kg and standard deviation 0.5 kg.

a. How often will the restaurant run out of coffee if they buy 5 kg each week?

b. What weight of coffee should you buy to be 99% sure you can serve all customers with coffee when it is requested?

c. What other assumptions have you made in your calculations?

12. Suppose that the time between ordering and receiving your order in the Quik Takeaways fast-food chain has been found to be normally distributed with a mean time of 10 minutes and a standard deviation of 2.5 minutes.

a. What is the probability that a randomly selected person will have to wait at least 6 minutes before his or her order is served?

b. In an advertising campaign, Raylene Brown, the manager of Quik Takeaways, offers a coupon for a free bottle of wine with the next purchase for all customers whose service time exceeds 10 minutes.

i. What proportion of customers would receive coupons if the process were improved so that the mean service time was reduced to 5.5 minutes, and the standard deviation remained at 2.5 minutes?

ii. Once the mean service time was reduced to 5.5 minutes, would Raylene be better off focusing on improvement that reduced the mean to 4.5 minutes or on improvement that reduced the standard deviation to 1.5 minutes. Explain the reasons for your answer.

13. In the Beckan Microchip Company, the average percentage defective expected from a process that produces Extel 967 microchips has been found to be around 3%. The company produces about 10,000 of these chips per week. How often will the weekly percentage defective exceed 3.5%?

14. The Statewide Insurance Company is monitoring the number of insurance claims. Over a number of years the average number of claims per quarter has been 7000. What is the probability that in the next quarter there will be over 7200 claims?

Chapter 5

Attribute Data

5.1 Analysing Attribute Data

You will recall from Chapter 3 that attribute data simply classify each item into one of a range of mutually exclusive categories. For example, a job meets specifications or not, a request is processed correctly or not, or a family belongs to socioeconomic group A, B, C1, C2, D, or E. Attribute data are often readily available from inspection records or market surveys and are usually quick and inexpensive to collect.

The only property that attribute data possess is that of equivalence (or not) between any two items. For example, two families either do or do not belong to the same socioeconomic group. As a consequence, the type of analysis that can be performed on attribute data is very limited. The only average that can legitimately be computed is the mode (the most frequently occurring category). Apart from that, analysis is usually confined to examination of the pattern of occurrence between the various categories.

To gain some insight into the sort of data we are concerned with and some of the issues that might arise, we begin by considering four examples.

⚒ *Absenteeism at Wellington Steel*

Fred Smith, the personnel manager of the Wellington Steel Company, is concerned about the level of absenteeism among the company's weekly paid workers. He chooses three representative weeks over the last 12 months and records the number of absentees for each day of the week, producing the data in Figure 5.1.

Monday	Tuesday	Wednesday	Thursday	Friday
39	16	28	26	31

Figure 5.1   Weekly absenteeism.

Q1.   What pattern seems to be suggested by the data? In particular, is there any evidence in the data of differences in absenteeism rates on different days of the week?

Q2.   What should Fred do about this problem?

📈 *Consumer Brand Preferences*

A toiletries and cosmetics company is introducing a new spray-on deodorant for both men and women. As a first stage in developing a marketing plan for the product, they conduct a small market research survey to examine customers' preference for the three leading brands of unisex deodorant. In a typical store, they observe a random sample of 150 purchasers of the three brands and record the brand chosen (A, B, or C) and the purchaser's sex. The results are given in Figure 5.2.

Q3.   What questions do you think would be of interest to the manufacturer of this product, and why?

Q4.   What do you conclude from the data?

Sex	Preference			Total
	Brand A	Brand B	Brand C	
Male	16	22	25	63
Female	25	40	22	87
Total	41	62	47	150

Figure 5.2   Results of market research survey.

Factory	Number of qualifying employees	Number accepting scheme
Anston	72	43
Butterfield	112	38
Carsbridge	56	27

Figure 5.3   Early retirement package acceptances.

🏭 Charnwood Casting Company

The Charnwood Casting Company has recently been exploring ways in which labour costs could be reduced and has decided to offer an early retirement package to all employees over 60. The company has three factories in the north of England, and a sample number of qualifying employees at each factory has been asked whether they are prepared to accept the package. The results, by factory, are shown in Figure 5.3.

Q5. In what way are the data in Figure 5.3 different from the data in Figure 5.2 (other than context)?

Q6. What are the two attribute variables observed in this example?

Q7. What questions would be of interest, and why?

Q8. What do you conclude from the data?

🏭 The Effectiveness of Bar Code Readers

A large supermarket chain is about to open eight new stores in the eastern region of the U.S. One of the many decisions to be made concerns the type of product scanning equipment that will be installed, and the company has to decide between two alternative types of bar code reader. To compare the effectiveness of the two readers, an experiment is conducted to determine how many times each reader cannot read the bar code on each of three types of surface: firm/curved (e.g., cans), firm/flat (e.g., packets), and flexible (e.g., bags). In the

Bar-code reader	Type of surface		
	firm/curved	firm/flat	flexible
Type A	24	13	40
Type B	15	18	22

Figure 5.4   Scanning errors on different surfaces.

experiment, 1500 different bar codes on each surface type are scanned using both readers, and the number of failures is recorded, as in Figure 5.4.

> *Q9. Does there seem to be a clear difference in the effectiveness of the two readers? If so, which appears to be more reliable?*
> *Q10. What other data might you want to consider before making a choice?*

5.2   Comparing Occurrences between Categories

We start by looking at the Wellington Steel absenteeism data in Figure 5.1. There are 140 absentees over the three-week period, and in each case the day of the week is recorded. So we have 140 observations of a single attribute variable (day) that has five categories. Clearly there are different numbers of absentees on each day of the week, but this is to be expected. As in the red beads experiment, we expect the results to vary to some extent, but is the variation here greater than we might expect if it were just down to chance? In other words, is there any evidence in the data of a systematic effect that causes more absentees on some days than we would expect? For example, the highest numbers occur on Monday and Friday. Is this just chance, or might there be something causing this? If we took another three weeks of data, would we get a similar "bias" toward Monday and Friday, or are we just as likely to get more absentees on Tuesday, Wednesday, or Thursday in the new set of data?

Let us assume that absences are essentially random events and in the long run can be expected to occur equally on any day of the week. If this were the case, how many occurrences of each day would you have expected to arise in the total of 140 observations? Write your answer in Figure 5.5, and compare these expected numbers with the observed values in Figure 5.1.

The question we have to answer is whether there is a real (*meaningful* or *significant*) difference between what we have actually observed in the sample (Figure 5.1) and what we would have expected if absentees on each day of the

Monday	Tuesday	Wednesday	Thursday	Friday

Figure 5.5   Expected numbers for different days.

week were, in fact, equally likely (Figure 5.5). For example, half of the total number of absentees occurred immediately before or after the weekend (70 out of 140). Is this observed number significantly different from the number we would expect, which from Figure 5.5 is 28 + 28 = 56? Think about the following two extreme situations:

- If our sample had contained 65 absentees on each of Monday and Friday and the remaining 10 spread over the other three days, this would surely be clear evidence of a general tendency for more absentees on Monday and Friday than on the other three days.
- If there had been 30 absentees on each of Monday and Friday, with 26 or 27 on the other three days, this would be close enough to an even split for us to conclude that the difference is due entirely to chance, that is, due to random variation.

So exactly where, in between these two extremes, do we begin to believe that the different days of the week are not occurring equally? As long as each expected number is reasonably large (say, greater than 5), then we can answer this question by computing a measure, called the *chi-square value*, of the difference between the actual and expected numbers. The procedure is as follows:

1. Square the difference between the actual and expected numbers for each category (i.e., day).
2. Divide each square by the corresponding expected number.
3. Sum the results for all categories to give the chi-square value. If this value is sufficiently large, the difference is *significant*.

The calculation for each day of the week is shown in Figure 5.6, giving a total of 9.929.

> **Q11.** *Why do we look at the squares of the differences rather than just the actual differences ?*

To decide whether the final figure of 9.929 is significantly large, we need to consult a table of the (so-called) *chi-square* distribution. This will tell us the probability of the sample result (the actual numbers) having arisen purely by chance if, in this case, absenteeism was equally likely to occur on any day of the week.

	Actual number	Expected number	Calculation
Monday	39	28	$(11)^2/28 = 4.321$
Tuesday	16	28	$(-12)^2/28 = 5.143$
Wednesday	28	28	$(0)^2/28 = 0.000$
Thursday	26	28	$(-2)^2/28 = 0.143$
Friday	31	28	$(3)^2/28 = 0.321$
Total	140	140	9.929

Figure 5.6   Calculation of the chi-square value.

A table of the chi-square distribution is given in Appendix A.4. It is used as follows:

1. First identify the number of *degrees of freedom* involved in your calculation. This is similar to the number of degrees of freedom in the calculation of the standard deviation, which you will recall was given by $n - 1$. In this case, it is not 1 less than the number of observations, but 1 less than the number of categories. In our example there are five days of the week, so we have $5 - 1 = 4$ degrees of freedom.
2. The number of degrees of freedom will identify a row of the chi-square table. Reading along this row, we see a set of increasing values, which are benchmark figures for our calculated measure (9.93). Each benchmark figure is associated with a probability, which is shown at the top of the column. So for 4 degrees of freedom, there is a 10% chance that it will exceed 7.78, a 5% chance that it will exceed 9.49, and so on.
3. Compare your calculated value to the relevant benchmark figures to reach a conclusion.

This procedure is called the *chi-square test* (for goodness of fit), but what does it tell us in this case? Our calculated value is just above the 5% benchmark figure. In other words, if absentees are equally likely on each day of the week, there is slightly less than a 5% chance of getting a random sample of 140 absentees that are as unequally split over the five days as those in Figure 5.1. Because 5% is a relatively low probability, it must cast some doubt on the supposition that absences on each day of the week are equally likely. Our result is said to be *significant at the 5% level.*

It is important to be clear about what this result means. It tells us that the days of the week are not all the same regarding absenteeism. However, we cannot necessarily infer from this result alone that Monday and Friday are different from Tuesday to Thursday, even if this pattern seems to be suggested in the data. If we wish to test this, we need to explicitly compare the two groups of days, Monday and Friday with Tuesday, Wednesday, and Thursday. We should be careful, however, not to be influenced in which comparisons to make purely on the basis of the data. There should be some *a priori* reason for the particular

Day	Actual number	Expected number	Calculation
Mon / Fri	70		
Tues / Wed / Thur	70		
Total	140	140	←

Figure 5.7   Chi-square calculations for the weekend effect.

comparison selected. In this case, a "weekend effect" is one that might be expected to occur, and it is plausible that Wellington Steel would be interested to examine whether there is significant evidence of this in the data.

The actual numbers are now as shown in Figure 5.7. If Monday and Friday were no different from any other day of the week, what would be the expected total number of absences on those two days? Write your answer in Figure 5.7, along with the expected total for Tuesday, Wednesday, and Thursday. Now complete the final column of Figure 5.7 to obtain the calculated chi-square value.

> *Q12. How many degrees of freedom are there in this case?*
> *Q13. What is your conclusion now about whether a "weekend effect" exists in the absenteeism figures?*

✍ *Using Excel*

The *chi-square* calculations can be done very simply in Excel. The probability that differences between what we observe and what we would expect are due to chance is given by a statistical function called *CHITEST*. This function depends only on the actual and expected numbers, which we have to enter in the spreadsheet. Note that when we use the table in Appendix A.4 we only know that the number we calculate is less than or exceeds a certain probability, or lies between two probabilities. *CHITEST* gives us the exact probability, which we can then use to judge whether the differences are real or not.

Figures 5.8, 5.9, and 5.10 show the relevant Excel windows used to examine the absenteeism data for a weekend effect. The data are entered as shown in Figure 5.8. The labels in column A and row 1 and the totals in row 4 are just for convenience; only the data in B2:C3 are strictly necessary. The probability in cell B6 is obtained from the function *CHITEST*. With the cursor on B6, select *Insert Function* to open the *Insert Function* window, as shown in Figure 5.9. Now enter the Actual and Expected ranges in the *CHITEST* window, as shown in Figure 5.10. The required probability (0.0157 in this case) is returned in cell B6.

	A	B	C
	Chi Test		
1	Day	Actual Number	Expected Number
2	Mon / Fri	70	56
3	Tue / Wed / Thur	70	84
4	Total	140	140
5			
6	Probability	0.0157	

Sheet1 / Sheet2 / Sheet3

Figure 5.8   Basic spreadsheet data.

Q14. *How do you interpret this probability of 0.0157?*

5.3   Excel Demonstration

To show how the results of a chi-square test depend on the pattern of the data, we have provided on the CD-ROM a demonstration of the how the *CHITEST* probability alters as the balance of the values in the frequency table changes. The spreadsheet, called *Chidem.xls,* comprises four sheets, the first of which,

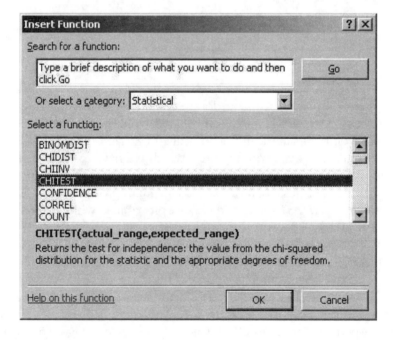

Figure 5.9   *CHITEST* function.

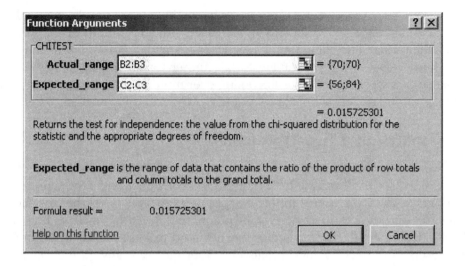

Figure 5.10 *CHITEST* **window.**

Wellington, relates to the analysis of the absenteeism data in the previous section. A copy of this sheet is shown in Figure 5.11. The other three sheets will be described later.

In Figure 5.11, the slider at the bottom left of the spreadsheet adjusts the balance between midweek absentees and Monday/Friday absentees. Moving the slider up creates more midweek absentees and fewer Monday/Friday absentees, with the values in cells C3 and C4 of the table adjusting automatically. As the balance gets closer to 84:56, the probability increases toward 1.0, indicating that

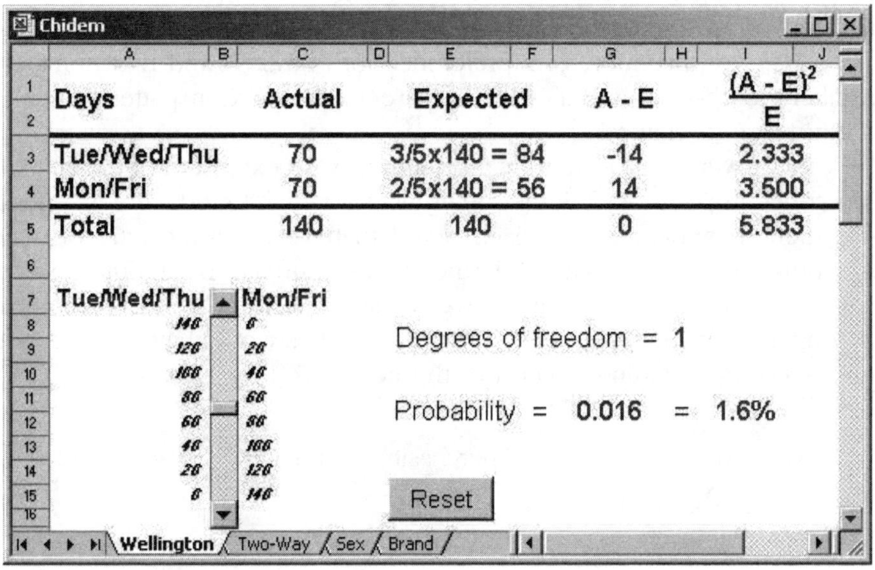

Figure 5.11   Demonstration spreadsheet for Wellington Steel.

the hypothesis of no weekend effect is becoming progressively more likely. Going beyond the 84:56 split, with an increasing number of midweek absentees, shows the probability falling until it reaches 0.016 again with a 98:42 split. Going even further causes the probability to fall further toward 0, indicating that the presence of a weekend effect is becoming ever more likely, but now with *fewer* Monday to Friday absentees than would be expected on a pro rata basis.

5.4 Analysing a Two-Way Table

The Wellington Steel example involved a single attribute variable, day of the week. In the second example in Section 5.1, on consumer preferences for unisex deodorants, data on two attribute variables were collected, the sex of the purchaser and the brand of deodorant purchased. We could analyse both variables independently to see whether females tend to be more, or less, frequent purchasers of the deodorants and also whether there are any differences in the preferences for the three brands. First, however, it is more appropriate to analyse both variables together. This will show whether males and females have different brand preferences. If they do not, then we can proceed to carry out independent analyses of sex and brand preference.

> *Q15. Why do you think the two-way table should be analysed before we consider each variable separately?*

A starting point for our analysis is, therefore, the two-way table of sex and brand given in Figure 5.2. There appears to be some evidence in the data to suggest that males and females have different preferences for these three brands. Females seem to have more of a preference for brands A and B, and males are more likely to choose C. But is this apparent difference in pattern significant, or is it just due to chance?

We can answer this question by a straightforward extension of the chi-square test described in Section 5.2. If the preferences for each brand were independent of sex, then we would expect each brand to have the same ratio of male to female purchasers. For instance, brand A was purchased 41 times. Since 63 among the 150 purchasers are male, we would expect males to have purchased a fraction 63/150 of these 41 items. Similarly, females would have been expected to purchase the remaining fraction of the items, 87/150. That is,

$$\text{Expected number of males purchasing brand A} = \frac{63}{150} \times 41 = 17.22$$

and

$$\text{Expected number of females purchasing brand A} = \frac{87}{150} \times 41 = 23.78$$

Sex	Preference			Total
	Brand A	Brand B	Brand C	
Male	17.22	26.04	19.74	63
Female	23.78	35.96	27.26	87
Total	41	62	47	150

Figure 5.12   Expected numbers for the two-way table.

Note that $17.22 + 23.78 = 41$, the total purchases of brand A. The full set of calculations for all three brands is given in Figure 5.12.

Q16. Verify the calculations for brands B and C.
Q17. What do you conclude from a comparison of Figures 5.2 and 5.12?

Since the row and column totals in Figure 5.12 must be equal to those in Figure 5.2, we have 2 degrees of freedom for the chi-square test. For example, once we have calculated the expected numbers of males for brands A and B (17.22 and 26.04, respectively), the remaining expected numbers are obtained by subtraction from the totals. In general the number of degrees of freedom in a two-way table is given by

$$\text{Degrees of freedom} = (\text{Number of row categories} - 1)$$
$$\times (\text{Number of column categories} - 1)$$

We can use the *CHITEST* function in Excel as before. The only difference is that the actual and expected numbers are now both two-dimensional arrays. This can be seen in the spreadsheet window shown in Figure 5.13.

	A	B	C	D	E	F	G	H
	Sex	Actual Number			Total	Expected Number		
2		Brand A	Brand B	Brand C		Brand A	Brand B	Brand C
3	Male	16	22	25	63	17.22	26.04	19.74
4	Female	25	40	22	87	23.78	35.96	27.26
5	Total	41	62	47	150			
6								
7	Probability	0.1615						

Chi Test — Sheet1 \ **Sheet2** / Sheet3

Figure 5.13   Spreadsheet for the two-way analysis.

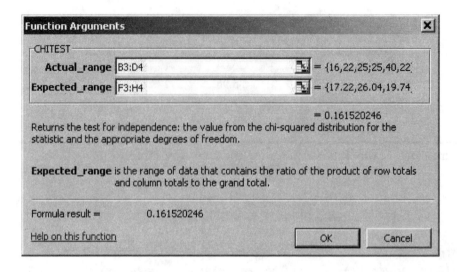

Figure 5.14   *CHITEST* window.

The expected numbers in Figure 5.13 can be calculated directly from the observed numbers in cells B3:D4, as follows. The total for males is calculated in cell E3 and the formula copied into cell E4. Similarly, the brand A total is obtained in cell B5 and the formula copied into cells C5:E5. The expected number in cell F4 is then given by the formula =$E3*B$5/E5, and this value is then copied into cells G3:H3 and F4:H4. Note the use of absolute cell references. (See Section 5 in the *Introduction to Excel* on the CD-ROM.)

The chi-square probability is determined (cell B6) using the *CHITEST* function, as shown in Figure 5.14.

> *Q18. How would you interpret the value of 0.1615 obtained?*

✍ *Excel Demonstration*

The Excel chi-square demonstration described in Section 5.3 has three further sheets that relate to the consumer brand preference data, one of which is a two-way analysis of sex and brand preference. A copy of this spreadsheet is given in Figure 5.15.

Figure 5.15 shows two sliders, because the table has two degrees of freedom. The number of male purchases of brand A and brand B can both be adjusted independently, with the male purchases of brand C and the female purchases of all three brands automatically adjusting to give the required row and column totals. Again, as the sliders are moved, the probability changes automatically between 0 and 1. As the pattern in the top and bottom rows of the *Actual Number* table becomes more unequal, the probability approaches 0, indicating a significantly different pattern of brand choice by males and females. Correspondingly, as

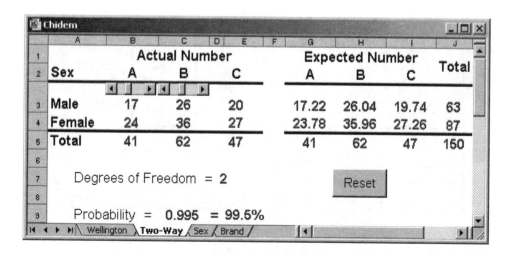

Figure 5.15 Demonstration spreadsheet for the two-way analysis.

the top and bottom row patterns become more alike, the probability approaches 1. For example, adjusting the male purchases of brands A and B to 17 and 26, respectively, gives a virtually identical purchase pattern for men and women, with a probability of 0.995, as shown in Figure 5.15.

> *Q19. If the data collected had shown purchasers of the three brands in four different age groups (under 25, 26–35, 35–50, over 50), how many sliders would be necessary to allow the balance of the table to be fully varied? With which categories of age/preference would they be associated?*

5.5 Looking at Margins

The analysis of the previous section is essentially examining whether there is a relationship between the two attribute variables, in particular, whether the choice of brand is related to the sex of the purchaser. The results of our analysis show that data such as that in Figure 5.2 are quite likely to occur when brand preference is unrelated to sex. The probability 0.1615 is quite large, indicating that sex and brand preference could well be independent variables. The chi-square probability would need to be much smaller (less than 0.05) for there to be clear evidence of a relationship.

Since it has been shown that there is no evidence to suggest that sex and brands are related, we can now proceed to examine each variable separately. To do this we carry out analyses similar to the one in Section 5.2 on the row totals and column totals of the data in Figure 5.2. That is, we analyse the margins of the two-way table. For sex, we see that there were 87 females and only 63 males.

	A	B	C	D	E	F
		| Chi Test				_ □ ×
1	Sex	Actual Number			Total	Expected
2		Brand A	Brand B	Brand C		
3	Male	16	22	25	63	75
4	Female	25	40	22	87	75
5	Total	41	62	47	150	
6	Expected	50	50	50		
7						
8	*Sex:*	Probability	0.0500			
9	*Brand:*	Probability	0.0963			

Sheet1 / Sheet2 \ **Sheet3**

Figure 5.16   Spreadsheet for margins analysis.

Is this evidence to conclude that women are more frequent purchasers of these brands than men, or are men and women equally likely? If they were equally likely, then we would expect 75 female and 75 males purchasers.

Q20. *Do you think females are more frequent purchasers of these brands than males?*
Q21. *If they are, in what way might the company make use of this knowledge?*

For brands, we see that brand A has been purchased 41 times, brand B 62 times, and brand C 47 times. Is it the case that the three brands are equally popular, or is there a significantly greater preference for brand B? With equal preference we would expect each brand to be purchased 50 times.

We can carry out chi-square tests in the same way as in Section 5.2. Alternatively, we can use the *CHITEST* function in Excel. First we set up a spreadsheet similar to that shown in Figure 5.16. Using the *CHITEST* function gives the probabilities in cells C8 and C9:

Sex: *CHITEST*(E3:E4,F3:F4) = 0.050
Brand preference: *CHITEST*(B5:D5,B6:D6) = 0.096

Q22. *What do you conclude from these CHITEST results? Do you think it is reasonable to suppose that, in general, purchasers are equally likely to be male or female and that the three brands are equally preferred?*

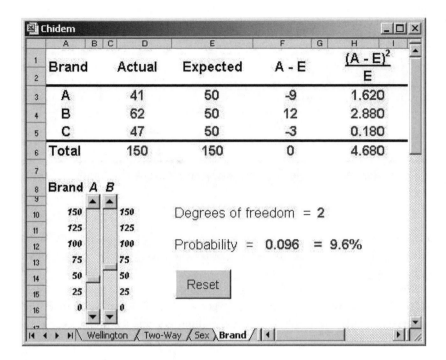

Figure 5.17   Demonstration spreadsheet for brand preferences.

Excel Demonstration

The final two sheets in the *Chidem* demonstration spreadsheet give independent analyses of both sex and brand preference. The margins analysis of sex is very similar to that for the absenteeism data at Wellington Steel. A copy of the brand preference sheet is shown in Figure 5.17. Figure 5.17 once again shows two sliders, because this table also has 2 degrees of freedom. The sliders change the numbers of purchases of brands A and B, with the number for brand C automatically adjusting to maintain a total of 150. The probability is 1 when the actual numbers are all the same; but as the balance becomes more unequal, the probability falls to 0. For example, a brand purchase pattern of 65:50:35 has a probability of about 0.01, meaning there is little chance this sort of pattern would occur if the brands were, in fact, equally purchased. In contrast, a pattern of 60:50:40 shows a reasonably high probability (0.135) and indicates no real difference in the numbers purchasing each brand.

5.6   Case Study: Charnwood Casting Company

As it stands, Figure 5.3, for the numbers of employees accepting Charnwood Casting Company's early retirement package, is not a two-way table. The counts in the last column are a subset of those in the previous column. The data recast

Factory	Number accepting package	Number not accepting package	Total
Anston	43	29	72
Butterfield	38	74	112
Carsbridge	27	29	56
Total	108	132	240

Figure 5.18   Two-way table for the early retirement package.

as a proper two-way table are given in Figure 5.18. Of interest in this example is whether the proportion of employees intending to accept the early retirement package is the same for each factory.

Entering the data of Figure 5.18 into a spreadsheet similar to that of Figure 5.13, the *CHITEST* probability is found to be 0.0024. This indicates quite conclusively that the three factories are unlikely to be equally receptive to the proposed package. For the data collected, the proportions choosing to accept are:

Anston: 60%
Butterfield: 34%
Carsbridge: 48%

These percentages appear to be significantly different, and this is born out by the chi-square test.

Q23. Verify that the CHITEST probability to answer this question is 0.0024.
Q24. Suggest two or three reasons why the factories may be exhibiting a different acceptance rate.

5.7   Case Study: Arcas Appliance Company

The Arcas Appliance Company is currently considering some changes to its pension scheme. Specifically, male employees' contributions are to be increased from 5% to 7% of the basic wage/salary. In return, pension provision for wives and children will be enhanced in the event the employee dies before reaching retirement age. It is envisaged that all new employees will be obliged to join the new scheme, but existing employees will be allowed to choose whether

Do you intend to change to the new pension scheme?	Age under 30		Age 30 or over	
	Married	Single	Married	Single
Yes	109	153	362	52
No	43	124	207	67
Don't Know	27	102	46	3

Figure 5.19   Acceptance of the pension scheme.

they will change from the old scheme. To assess the likely level of acceptance of the new scheme by existing employees, a survey was conducted at one of the company's factories. The results in Figure 5.19 are tabulated according to age and marital status.

Q25. What questions would be of interest, and why?
Q26. What tentative conclusions do you draw from the data?

Before making any changes Arcas wishes to know whether acceptance of the scheme is influenced by either age or marital status. For example, it is plausible that unmarried employees will not be as strongly attracted to the scheme as those with dependents, for whom provision has to be made in the event of the employee's early death. Likewise, younger employees may not want to sacrifice earnings over many years for potential benefits a long time ahead. These effects may be apparent in the data, but are they significant?

One of the key issues in answering these questions is how to deal with the *Don't Knows*. If they are treated as a separate category for analysis purposes, it could be that a significant result is obtained mainly due to the effect of the *Don't Knows*. Ultimately, all current employees must decide whether to change or not, so should the *Don't Knows* just be disregarded? One of the drawbacks to this is that they constitute almost 14% of the people in the sample, and, because they must eventually be either *Yes* or *No*, maybe we should try and anticipate how they will eventually split on the basis of the present balance of *Yes* and *No* in each of the four age/marital status groups. There are, therefore, at least three ways of proceeding:

1. Ignore the *Don't Knows*.
2. Keep the *Don't Knows* as a separate category.
3. Allocate the *Don't Knows* pro rata to the *Yes* and *No* categories.

Do you intend to change to the new pension scheme?	Age		Status	
	Under 30	30 or over	Married	Single
Yes	262	414	471	205
No	167	274	250	191

Figure 5.20   Acceptance by age and marital status.

> **Q27.** *Which alternative would you choose, and why?*
> **Q28.** *If you choose to allocate them pro rata, how many of the 27 "Don't Knows" that are married and under 30 would be allocated to the "Yes" group and how many as "No"?*

For simplicity, suppose we choose to ignore the *Don't Knows*. We can create two separate two-way tables showing:

- Choice against age (with the numbers summed over marital status)
- Choice against marital status (with the numbers summed over the two age groups)

This gives the tables in Figure 5.20.

Each table has 1 degree of freedom, and using the *CHITEST* function gives probabilities of 0.7653 for choice against age and 0.000009 for choice against marital status. The results are therefore very clear-cut. There is no evidence of any relationship between choice and age. About 60% of employees have chosen to change to the new scheme, and this proportion is true for both those under 30 and those over 30. However, a highly significantly different pattern of choice exists for married versus single employees. About 65% of married employees have chosen to change to the new scheme, but the proportion is only about 52% within the single group.

However, we must be careful when analysing a three-way table such as that in Figure 5.19. Looking at whether age and marital status independently influence a person's choice may be misleading because it does not take account of any possible relationship between age and marital status. In this case such a relationship exists because there are proportionately more married people over 30 and more single people under 30. One consequence of this is that even though we see no overall relationship between age and choice, there *is* some evidence of a relationship if we consider married and single people separately. In particular, for both marital status groups, proportionately more people in the under 30 group choose the new scheme than in the over 30 group.

5.8   Chapter Summary

This chapter has been concerned with analysing the patterns that occur in distributions of attribute data. Specifically:

- The use of one-way tables to describe the distribution of a single-variable and two-way tables for joint distributions of two variables
- How the pattern in a one-way table can be compared with some expected distribution
- Use of the chi-square test to determine whether a significant difference exists between observed and expected values
- How to determine the degrees of freedom in both one- and two-way tables
- Use of the chi-square distribution to determine a "significance probability"
- How the expected frequencies are determined in a two-way table
- Use of the chi-square test to determine whether a significant relationship exists between two attribute variables
- How the margins of a two-way table can be analysed as individual attribute variables if the two variables are independent.
- Different ways in which *Don't Knows* can be dealt with in survey data

5.9   Exercises

1. The management of a store is interested in whether the method of payment used is different between weekends and weekdays. The data in Figure 5.21 were collected during a particular week. From these data the *CHITEST* probability is 0.083. Is there is any clear evidence that the method of payment used is different on weekends than on weekdays? Write a few sentences to describe the pattern of usage of different payment methods on weekdays and on weekends.

2. A mail order company uses two different couriers, NTN Logistics and Packetforce, to deliver customer orders. The contracts with both couriers are due for renewal, and you have been asked to investigate their performance. One measure of performance is the proportion of orders delivered within the specified time. A sample of orders delivered by each courier is analysed, yielding the data in Figure 5.22.

    a. Is there any evidence in the data that one company is performing better than the other in terms of on-time deliveries?

	Cash	Cheque	Charge card	Credit card
Weekday	327	55	295	429
Weekend	239	44	287	356

Figure 5.21   Method of payment.

Courier	Number of orders delivered	Number delivered on time
NTN	452	357
Packetforce	626	466

Figure 5.22   Orders delivered on time.

   b. The contract with each company states that the company should achieve
       at least 80% of deliveries within the specified time. Is there any evidence
       that either of the companies is failing to achieve this target?
3. It has been suggested that the public's confidence in big business is closely
   tied to the economic climate. When business is contracting and unemployment
   is increasing, public confidence is low. When the opposite occurs, public
   confidence is high. In a market research study, a random sample of 810 people
   in employment were asked questions about their job security and their degree
   of confidence in big business. The results are given in Figure 5.23.
   a. Explain carefully what you regard as the most important question that
       should be asked.
   b. The corresponding expected numbers are given in Figure 5.24. Explain
       precisely what the data in this table indicate. Verify that the value in the
       first cell is 102.0.
   c. From the two tables, a chi-square value of 14.55 is calculated. How many
       degrees of freedom are associated with this value?
   d. What conclusions do you draw from this analysis? Relate your conclu-
       sions to the suggestion made in the first sentence of the question.
4. The Llundain Tool Company employs three account clerks, Albert, Bernard,
   and Colin. Recent concerns about the number of incorrectly prepared invoices
   have caused the accounting manager to examine the clerks' accuracy. The
   number of invoices prepared by each clerk over the last two weeks and the
   number in error are given in Figure 5.25.
   a. There are two attribute variables. What are they?
   b. If the clerks had an equal "error rate," how many invoice errors would
       you expect for each clerk?
   c. Set out the data in a two-way table. Is the difference between what occurred
       and what you expect on the basis of an equal error rate significant?

Confidence in Big Business	Job security			
	Very secure	Moderately secure	Slightly insecure	Very insecure
A great deal	120	53	10	6
Only some	255	168	41	22
Hardly any	62	47	15	11

Figure 5.23   Observed job security and business confidence.

Confidence in Big Business	Job security			
	Very secure	Moderately secure	Slightly insecure	Very insecure
A great deal	102.0	62.5	15.4	9.1
Only some	262.2	160.8	39.6	23.4
Hardly any	72.8	44.7	11.0	6.5

Figure 5.24   Expected job security and business confidence.

	Albert	Bernard	Colin
Invoices prepared	164	219	217
Errors involved	13	21	14

Figure 5.25   Errors in invoices.

5. In a market research survey concerned with home loans, a random sample of 262 people who had taken out a first mortgage in the last 12 months was interviewed. From the interviews, the data in Figure 5.26 were collected.
    a. Does there appear to be any difference between the size "profile" of loans made by banks and building societies?
    b. Use the chi-square test to examine this formally.
    c. Does it make a difference to your conclusions if the comparison were made simply between loans of less than or equal to £50,000 and those more than £50,000?
    d. Write a short paragraph describing your findings.
6. A management accountant is investigating the budgetary control procedures used in her company. She takes a random sample of 85 projects undertaken during the last five years and assembles the information in Figure 5.27 regarding the nature of the project and the variance from budget.
    a. Is there any evidence in the data of more effective budgetary control in certain types of projects?
    b. Briefly comment on your findings.
7. A management accountant is asked to investigate the effectiveness of stock control procedures in use in a large automated warehouse. The warehouse

Source of loan	Size of loan		
	< £20,000	£20,000 - £50,000	> £50,000
Bank	10	14	25
Building Society	54	83	76

Figure 5.26   First mortgages.

Project type	Variance from budget		
	Below budget	0 – 10% above	>10% above
Capital	2	16	4
R&D	0	13	24
Other	1	17	8

Figure 5.27   Budgetary control in projects.

Item Category	Number of stock-outs		
	None	1 - 3	4 - 10
A	5	11	9
B	10	29	16
C	12	23	35

Figure 5.28   Stock-outs according to item category.

holds a wide range of stock items that are classified according to the value of annual usage as:

Category A   High usage (>$1000)
Category B   Medium usage ($100–$1000)
Category C   Low usage (<$100)

A random sample of 150 stock items is taken, and the number of stock-outs of each item during the previous 3 years is recorded, giving the data in Figure 5.28.

a. Is there any evidence in the data of a different pattern of stock-outs in the three item categories?

b. Estimate the average number of stock-outs *per year* for each item category, and comment briefly on your results.

8. As part of a research project into capital investment appraisal techniques, a researcher surveyed 193 companies to determine which technique they *mainly* used to assess the viability of major capital investment decisions. Tabulating the technique used against the size of the company gives the data in Figure 5.29.

Technique	Size of company (thousands of employees)		
	Large (> 10)	Medium (1 – 10)	Small (< 1)
NPV/IRR	27	25	8
Payback period	16	39	15
Other	13	26	24

Figure 5.29   Investment appraisal techniques used according to company size.

Advertisement	Number watching TV	Number aware of advertisement
A	228	155
B	216	121

Figure 5.30   Advertising awareness data.

    a. Is there any evidence in the data that the technique used is influenced by the size of the company?

    b. Write a brief report describing the use of alternative capital investment appraisal techniques in different size companies.

9. An advertising agency has been commissioned to develop a TV advertising campaign for a major bank. The campaign involves two new advertisements appearing at peak viewing times between 7:30 and 8:00 each evening. After the first screening of each advertisement, the agency surveys a viewers' panel of 386 randomly chosen people to determine whether they were watching TV at the time and if they can recall the name of the bank being advertised. They obtained the results given in Figure 5.30.

    a. Compare the viewers' awareness of each advertisement.

    b. A chi-square test is carried out to determine whether there is any real evidence that one of the advertisements has had more impact than the other. The calculated chi-square value is 6.75. Explain carefully how this value is obtained, calculating any expected numbers required. What do you conclude?

10. The management of Muchsave Stores is concerned that too few customers are using the available hand scanners, causing longer lines at the checkouts. A particular store is chosen and all customers entering the store on Tuesday morning are observed to see whether they choose to use a hand scanner. The results are given in Figure 5.31.

    a. What pattern appears to be present in the data?

    b. To confirm the apparent pattern, a chi-square test is performed. What is the expected number of men who would use a hand scanner? Hence, deduce the other three expected values.

    c. Calculate the chi-square value for the data.

    d. Write a short paragraph stating your conclusions about the pattern of use of hand scanners by men and women.

Used a hand scanner	Male	Female
Yes	27	35
No	88	372

Figure 5.31   Use of hand scanners.

Chapter 6

Sampling

6.1 Collecting Data

In Chapter 3 we looked at different types of data and the reasons we collect data. In this chapter we consider how to gather data. Most organisations collect a variety of data relevant to their operations. Companies are likely to collect sales data, customer data, and data about their employees. Universities collect data about students and their performance. National governments collect data about a vast array of things, such as age distribution, the country's trading and business activity, and its economic performance. We begin by considering four examples of data collection.

Should New Zealand Become a Republic?

In August 2000, the *New Zealand Herald* newspaper invited its readers to participate in a poll about whether New Zealand should become a republic. Readers were asked to call a 0900 number, at a cost of about $1 a minute, and register their opinion. After the poll, they published the results given in Figure 6.1.

> **Q1.** *Do you regard this as a representative sample of the New Zealand population? Why or why not?*
>
> **Q2.** *Is it reasonable to conclude that the majority of the New Zealand population are against a republic?*

Population Census

Many countries aim to measure the size, composition, and characteristics of their population on a regular basis by conducting a population census, usually

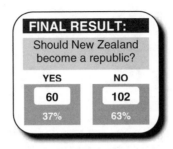

Figure 6.1   Republic of New Zealand?

every 10 years. For example, the U.S., the U.K., and New Zealand each conducted their most recent population census in either 2000 or 2001.

> *Q3.   Why is a population census important? Are there any disadvantages of such a census?*
>
> *Q4.   Is there any other way of reliably obtaining the same information?*

📊 Superannuation Referendum

In 1997 a referendum was carried out in New Zealand on the issue of compulsory superannuation. In essence, under such a scheme, all adults will be obliged by law to make financial provision for their retirement by making regular contributions throughout their working life into an approved pension fund. A postal ballot of all people on the Electoral Register was carried out. They were asked whether they were for or against a compulsory superannuation scheme.

> *Q5.   What are the disadvantages of such a referendum?*
>
> *Q6.   Can you think of an alternative method of collecting the information? What are the advantages and disadvantages of your method?*

📊 Market Research Survey

A sculpture exhibition was held in an enclosed area in Battersea Park in London. The organisers wanted to sample people attending the exhibition to determine who is attending, where they come from, and how they found out about the exhibition. This information will be useful in planning and marketing future exhibitions.

> ***Q7.*** *Write down two ways in which such a sample might be taken. What are the relative merits and disadvantages of your methods?*

6.2 Sampling

Whenever we collect data, we usually have to sample the process or situation under study. In the superannuation and market research examples, the data required were, or could have been, collected by taking a selection from the population concerned. Any selection or subset of some population is called a *sample*. Sampling is very often the only practical way of obtaining data about large populations in a cost effective and timely way.

Suppose in the superannuation referendum we decided to estimate the proportion favoring some form of compulsory superannuation by taking a sample of eligible voters.

> ***Q8.*** *What factors do you think will influence the choice of the size of the sample?*
> ***Q9.*** *How would you choose a sample of, say, 5000 voters?*
> ***Q10.*** *What are the advantages and disadvantages of sampling in this case?*

✎ *Sampling in Practice*

Sampling is used in most areas of business, economic, social, and political activity. Here are some examples:

- *Monitoring industrial processes*. As an illustration, an injection-moulding machine is making plastic pipes. In order to check whether the machine is producing on target, the internal diameter of a sample of five pipes is measured every hour.
- *Auditing invoices*. Auditors sample invoices to check the level of errors. They may aim to check the higher dollar-valued invoices more thoroughly, where there is more scope for large errors.
- *Product design*. A company is about to develop a new style of DVD player. It samples current and prospective customers to find out which brands they

are already aware of and which aspects of their appearance and performance are most important to them.

■ *Political opinion polls.* These polls sample the population of registered electors who are regularly questioned to obtain their preferences for political parties.

■ *Environmental monitoring.* Waste is treated before being discharged into a stream. Because fines are imposed for high levels of contaminants, it is important that these not exceed the legal limits. To monitor this, regular samples of stream water are taken to check the level of contaminants.

■ *Market research.* Most organisations consider it important to have detailed knowledge of their customer base. They frequently conduct surveys to find out why customers bought their products or services. This is often done by using representative samples of customers to form customer panels and focus groups.

6.3 Sampling Concepts

Sampling is used to get information about a population or process. We use a sample because it is often too expensive, time consuming, practically impossible, or simply unnecessary to canvass the whole population. The following are some examples:

■ A census or referendum is expensive. The required information can usually be obtained from a sample. Further, it is often more accurate to take samples, since the resources available can be focused on the smaller sample rather than across the whole population.

■ It is time consuming and costly to inspect every invoice raised or every item produced.

■ Quick results may be needed on an important issue, such as a decline in sales figures.

■ It is impossible to study the whole population if the sampling method is destructive, for example, when cutting fruit in half to measure fruit quality, when a can of baked beans is punctured to measure the bacterial count, or when the lifetime of an electric light bulb is required.

In order to facilitate a discussion of sampling methods we need to define some terms.

A *population* is the set of all the individuals or items in a group of interest:

■ This may be a finite collection, such as the population of students enrolled in a particular course.

■ It may be an infinite collection or a very large group, such as the population of $2 coins in circulation.

A *sample* is a subset of the population, for example, a subset of the students in a course that form a tutorial group, the voters living in a certain area of a parliamentary constituency, or the first hour's production of extruded items in a plastics factory.

A *parameter* is a measurable characteristic of the *population,* for example, the population mean or the population standard deviation.

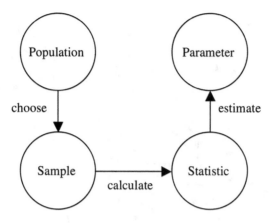

Figure 6.2   Sampling process.

A *statistic* is a measurable characteristic of the *sample,* for example, the sample mean or the sample standard deviation.

In principle any subset of the population is a sample, no matter how it is selected. In practice, however, we need to select the sample carefully so that it is representative of the population from which it is taken. If a sample is properly selected, particular population characteristics should be reflected in the corresponding sample characteristics. In contrast, a sample that is not properly selected is likely to be nonrepresentative and to give a distorted picture of the population.

The aim of sampling is to *choose* a subset of the population from which we *calculate* appropriate sample statistics that accurately *estimate* population parameters. This *choose–calculate–estimate* process is depicted in Figure 6.2.

In Chapter 4 we saw how to calculate various statistics, such as the mean and standard deviation, from a sample of data. In the next chapter we will consider some of the issues involved in estimating population parameters. In the following sections of this chapter we will look at the key considerations we should look at when choosing a sample of data and describe different ways in which the sample could be chosen.

6.4   Wood Experiment

We can select a sample from a population in a number of ways. We discuss the more frequently used sampling methods in Sections 6.5 and 6.6. Before doing that, however, we introduce some of the key issues in the choice of a sample by means of a simple experiment: the wood experiment.

The experiment involves estimating the mean weight of a population of 100 blocks of wood of varying shapes and sizes but of the same thickness. The blocks range in weight from a minimum of 5 g to a maximum of 105 g, with weights recorded to the nearest 5 g. A pictorial representation of these 100 blocks is given in Figure 6.3.

We could, of course, calculate the mean of all 100 blocks to determine the *population* mean. However, for the purpose of this experiment, we will suppose

Figure 6.3   Population of 100 blocks of wood.

that the population mean is unknown and that we wish to estimate it from a sample of 10 blocks.

We shall use two different methods of sampling:

- A judgement sample
- A random sample

We will describe various procedures for both random and judgement sampling in more detail in Sections 6.5 and 6.6. For the present, we will say simply that a *judgement sample* is one where a person uses his or her best judgement to select a sample he or she considers representative of the population being sampled. In contrast, a *random sample* is chosen completely objectively, without any reference to the population being sampled.

> **Q11.** *What is your instinctive assessment of these two sampling methods? In particular, which method will be more accurate in estimating the mean weight of the population?*

✍ *Judgement Sample*

First use your judgement to select a sample that you think will give an accurate estimate of the mean weight of the population. To do this:

1. Look carefully at the 100 shapes in Figure 6.3. Choose a sample of 10 of these that you regard as representative of the population. Write their numbers in the *Block number* column in Figure 6.4.

	Block number	Weight
1		
2		
3		
4		
5		
6		
7		
8		
9		
10		
Mean weight		

Figure 6.4   Judgement sample results.

2. The weights of all 100 blocks are given in Figure 6.9 (in Section 6.9). Using these data, write the *Weight* of your chosen blocks in Figure 6.4.
3. Calculate the mean weight.

✍ *Random Sample*

Now select a second sample, using a random approach. To do this use the random number table in Appendix A.5 as follows.

1. It is necessary to choose a starting point in Appendix A.5 at random. One way to do this is by using your birthday. The day gives the row and the month the column. For example, if your birthday is 27 March, then start with the number in the 27th row and 3rd column; that is, 67. Write this number and the next nine numbers, working down the column (go to the top of the next column if necessary) in the *Block number* column in Figure 6.5.
2. Again using the data in Figure 6.9, write the *Weight* of your chosen blocks in Figure 6.5.
3. Calculate the mean weight.

> **Q12.** *The population mean weight of all 100 blocks is 39.4 g. Which of your estimates is closer to this population mean? Are you in any way surprised by your two results?*

If you are able to take part in the wood experiment with a group of students, construct a back-to-back stem and leaf plot (similar to that in Figure 6.6) of the results obtained by everyone in the group.

	Block number	Weight
1		
2		
3		
4		
5		
6		
7		
8		
9		
10		
	Mean weight	

Figure 6.5   Random sample results.

Random		Judgement
Leaf	Stem	Leaf
	1	88
	2	
	2	2
45555	2	
77	2	7
8889999	2	89999
00011111111111	3	011
222333333	3	233
44444445555555555555	3	44445555555
66666666666666677777777777777	3	66666666677777
888888888888888888999999999999999999999	3	8888888888999999
0000000000000000111111111111	4	0000000000000000011111111111111
2222222222222223333333333333333333	4	222222222223333333333333
44444444444444455555555	4	44444444444444455555555555555
66666777777	4	666666666666677777
8888889999	4	88888888889999999999
0000001111	5	0000000000111111111111
23	5	2222222233333
4	5	4455555
	5	6666666677
8	5	8899
	6	01
	6	22
	6	455
	6	
	6	9
	7	01
	7	2

Figure 6.6 Stem and leaf plot of wood experiment results.

To illustrate the "big picture" when the experiment is carried out with a large group, Figures 6.6 and 6.7 show a set of results from a class of 243 students. These students actually inspected a population of real blocks of wood, whose weights were the same as the values in Figure 6.9. Each student selected two samples exactly as you have done and calculated the mean weight of each sample. The results of the experiment are given as a back-to-back stem and leaf plot in Figure 6.6 and as a box plot in Figure 6.7.

Using either your stem and leaf plot, or the results of the group of 243 students summarised in Figures 6.6 and 6.7, answer the following five questions.

Q13. Roughly estimate the mean of each distribution.
Q14. What can you say about the amount of variation in the results from the two sets of samples?
Q15. Which sampling method is better in estimating the population mean, and why?

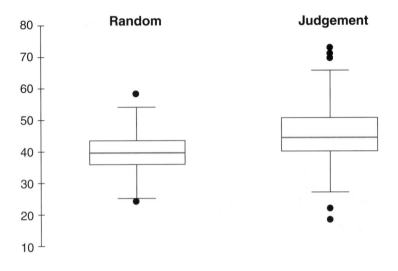

Figure 6.7   Box plot of wood experiment results.

> ***Q16.*** *The judgement sample tends to overestimate the population mean, as can be seen in Figure 6.7. Can you think of a reason why this is the case?*
>
> ***Q17.*** *What are your overall conclusions from this experiment?*

There are three main conclusions that emerge from the data in Figures 6.6 and 6.7.

1. The judgement sample is clearly giving biased results, since the average of the judgement means is greater than the population mean of 39.4. The mean of the 243 judgement sample means is in fact 44.7, whereas the mean of the 243 random sample means is almost exactly 39.4. Despite using our best judgement to obtain a sample that is representative of the population, we are overestimating the population mean by about 5 g. The reason for this is simple. When we inspect the population in Figure 6.3, our eyes are drawn to the relatively few large blocks, and we don't notice the many smaller blocks. Hence, we tend to overrepresent the large blocks in our sample.

2. There is clearly more variability in the judgement sample means as compared to the random sample means. From the box plot in Figure 6.7, the inter-quartile range for the judgement samples is 10, whereas that for the random samples is 8.5. The random sample means are more consistent and so produce more accurate estimates of the population mean, even if the judgement samples were not biased.

3. The distribution for the random sample means seems to resemble very closely that of a normal distribution, whereas that for the judgement sample means is more skewed. In fact, as we pointed out in Chapter 4, the distribution of a (random) sample mean will get ever closer to a normal distribution as the sample size increases. Even with samples of size 10, the normality is clearly evident. In contrast, no predictable distribution occurs with the judgement sample means. As a consequence of this, as we shall see in Chapter 7, it is possible to predict the likely accuracy of a mean obtained from a random sample but not one based on a judgement sample.

6.5 Random Sampling

A random sample is one that is selected objectively by some random process, where there is no element of judgement on the part of the sampler over which particular items are chosen from the population. Random samples are sometimes referred to as *probability* samples because this method of selection implies that the probability that an individual item or a group of items will be selected is known.

> *Q18. Consider the example at the beginning of this chapter concerning whether New Zealand should become a republic. Is the sample of responses a random one? Why, or why not?*
>
> *Q19. Is the sample of responses obtained in the superannuation referendum a random one? Why, or why not?*

Random samples can be obtained in a number of specific ways.

✍ *Simple Random Sampling*

The random sample used in the wood experiment is an example of a simple random sample. In such a sample, each possible sample of a given size is equally likely, or has the same chance as any other, of being selected. This is one of the most common probability-based sampling schemes. In order to take such a sample, we must:

■ Have a list of the members of the population; this is called a *sampling frame*. For example, the electoral roll is a sampling frame for the voters in an electorate. Upcoming Figure 6.9 gives a sampling frame for the wood experiment.

- Decide on the sample size. A method for doing this is discussed in detail in Chapter 7.
- Choose items from the sampling frame using some random process, for example, a table of random numbers such as the one in Appendix A.5. Each group of items of the chosen size then has the same chance of being selected for the sample.

✥ Stratified Sampling

In a stratified sample, the population is first divided into natural groups, or *strata*. Then a simple random sample is taken from each stratum. The number sampled from each stratum is often chosen to be proportional to the total number in the stratum. Additionally, the sample should take into account the variability in each stratum. For instance, if the members of a stratum are relatively homogeneous, then only a small sample is required. A more variable stratum, however, requires a larger sample to give an accurate representation of the members of that stratum. Stratified sampling gives more precise estimates of population parameters than a simple random sample of the same size when the strata differ with respect to the parameter being estimated. However, to use stratified sampling effectively, we must have some knowledge of the composition of each stratum, such as its size and variability.

To illustrate the process of stratified sampling, suppose we wish to determine the average amount of money spent each week by undergraduate students at a university. The university has 15,000 students, 40% of whom are in the first year, 30% in the second year, and 30% in the third year. In each year 60% are female and 40% are male. If we suspect that the amount of money spent is likely to differ between male and female students but not between different years, then we will generally improve the accuracy of our estimate of the average amount spent by sampling male and female students independently. Given the composition of the student population, we would allocate 60% of the sample to female students and 40% to male. If we decide that a sample of 200 students will be adequate for our purposes, then 120 of the students sampled should be female and 80 male.

The preceding calculation is based on the assumption that male students' expenditure is just as variable as that of female students. But suppose we suspect that the variability of the men's spending is 50% more than that of the women. We need to build this differential variability into our determination of the sample allocation. To do this, we must weight the female proportion by 1 and the male proportion by 1.5 so that the allocation weight is $60 \times 1 = 60$ for the females and $40 \times 1.5 = 60$ for the males. The result is that our sample of 200 students should now by equally divided between males and females, i.e., 100 of each.

Suppose, however, that expenditures differ across the three student years. This implies that we now need to consider six strata rather than just two, namely, male and female students in each of the three years. Without involving the variability factor, the number of students sampled from each of the six strata is shown in Figure 6.8. For example, $40\% \times 60\% = 24\%$ of the sample, i.e., 48

Sex	Undergraduate Year		
	First (40%)	Second (30%)	Third (30%)
Male (40%)	32	24	24
Female (60%)	48	36	36

Figure 6.8 Stratum allocation considering both sex and year.

students, should be first-year females and 30% × 40% = 12%, i.e., 24 students, should be third-year males.

If we allow for the differential variability in male and female students' expenditure, as before there should be equal numbers of male and female students, so we need 100 males and 100 females, each split 40, 30, and 30 between the three years.

Q20. Why have we taken a stratified sample in this case? Why not a simple random sample?

✥ Systematic Sampling

Suppose we have a list of the items in the sampling frame. With systematic sampling we choose items at regular intervals from the list, such as every fifth or eighth item, with the starting number chosen using random numbers. For example, if our sampling frame is a telephone directory, then we might choose every 80th entry in the book, starting at the top of the second column on page 43. If the items in the sampling frame are listed in random order, systematic sampling is as good as simple random sampling, but it is generally much easier to administer. Once you have located the starting point, the procedure is straightforward.

A potential problem with this type of sampling is *periodicity*. That is, if there is a cycle in the data, we can get badly biased estimates, especially if the cycle has the same periodicity as the sampling interval. For example, suppose we sample every seventh day to estimate yearly sales at a retail outlet. If the pattern of sales differs from one day of the week to another, then our estimates could be seriously biased.

Q21. Returning to our example on sampling students at a university to determine the amount of money spent, how would you take a systematic sample of 200 students?

✍ *Cluster Sampling*

In stratified sampling a simple random sample is taken from each stratum. However, in practice it is sometimes impractical to take samples from every stratum, especially if there is a large number of them or they are widely dispersed geographically. Instead it becomes more sensible first to select a few of the strata at random and then to take a census or a random sample from each of the selected strata. This type of sampling is called *cluster sampling;* to avoid confusion between this and stratified sampling, we call the natural groupings of the population *clusters* rather than strata. For example, the clusters may be geographical areas, students on different courses, or days of the week.

Another important difference between stratified and cluster sampling is the composition of the strata or clusters. For stratified sampling it is desirable that the members of a stratum be relatively homogeneous. In contrast, for cluster sampling to be effective, each cluster should be representative of the population as a whole, because, typically, only a small number of clusters will be used for the sample. Cluster sampling is likely to give inaccurate results if the clusters are homogeneous. For example, if clusters are streets within a city, then basing a political opinion poll on a few randomly selected streets could be unwise if the households within a street have similar political views.

One of the main advantages of cluster sampling, other than economy and convenience, is that it eliminates any need for a sampling frame of the entire population. We simply need a list of the clusters, so we can identify all of the clusters as a basis for choosing one or more of these to constitute the required sample. As an example, suppose we need to check the quality of items being manufactured in a large machine shop comprising 84 similar machines. If we require a 5% sample of a week's output, a very simple and convenient way to get the sample is to take the entire week's output of a few of the machines. In this case each machine constitutes a cluster; if the machines are producing at more or less the same rate, then four or five machines (i.e., about 0.05×84 machines) should be selected at random to give the required sample.

> **Q22.** *Returning again to our example on sampling students, how could you take a cluster sample of 200 students? What is the potential danger of cluster sampling in this situation?*

6.6 Nonrandom Sampling

Nonrandom sampling occurs when the process of sample selection is not completely objective and where some form of judgement or subjectivity is involved. Even though it is advisable to take random samples wherever possible, nonrandom samples are often unavoidable. *It is important to realize, however, that assertions about the population based on nonrandom samples may be invalid.*

For example, we saw in Figures 6.6 and 6.7 that judgement sampling led to biased estimates of the mean weight of the wood population. Generally, experience has shown that judgement samples often tend to be biased upward.

> **Q23.** *Explain why the sample selected in the "Should New Zealand Become a Republic?" survey (Section 6.1) is a nonrandom one.*

We now briefly consider three particular methods of nonrandom sampling, although in practice nonrandom samples may be selected in many other ways in particular situations.

✍ Quota Sampling

A quota sample is one that is representative of various factors such as age, sex, and socioeconomic status. The person taking the sample has freedom to choose a specified number (or quota) of people fitting a certain profile, such as a given sex, age, and social class. Quota sampling is essentially the nonrandom counterpart to stratified sampling, where various groups (strata) are judgementally, rather than randomly, sampled.

Quota sampling is often used in market research surveys. The advantage of this method is that it is simple and relatively inexpensive to carry out. However, being nonrandom, the method can give an unrepresentative sample since the interviewer's personal biases, conscious or unconscious, influence the selection of sample members.

> **Q24.** *A market researcher is sampling customers leaving an up-market retail store. Her quota consists of interviewing a certain proportion of males and females of different ages and employment status. The profile of the selected subject is determined from the answers to the first few questions. Why might her results be biased?*

✍ Judgement Sampling

Here we do our conscious best to get a "fair" sample by picking those items that, in our judgement, form a representative cross section of the population. However, it is almost impossible for any person to be totally objective, so the risk of distortion due to the sampler's personal prejudices is serious, as the wood experiment clearly demonstrated.

✍ *Self-Selected Sampling*

In self-selected sampling, inclusion in the sample is left up to the members of the population, each of whom decides whether to be involved. Only those respondents sufficiently motivated to reply, or who choose to take part, are included. This is often seen on "talkback" TV or radio programmes, where the viewer or listener is asked to phone in about an area of topical interest. This method encourages people with strong opinions to participate, and for this reason it is unlikely that the sample will be representative of those with more moderate opinions.

> **Q25.** *Over 4000 people called into a TV talk show to express their vote for or against changing the name Hamilton to Waikato City. Over 80% were opposed to the change. What conclusions do you draw from this survey?*

One of the less obvious situations where self-selected sampling arises is where information is collected by means of a questionnaire. In most cases, completion of a survey questionnaire is voluntary, and the people who participate are doing so from personal choice. In other words they are self-selected respondents. A major issue here is whether those who choose to participate can be considered representative of those who do not, or is there a respondent bias?

6.7 The Accuracy of Sample Estimates

There are many reasons why the data produced by a sample could be inaccurate, even worthless. Even if a random sample has been used, the sample may not be representative of the entire population simply on account of the chance element involved in random sampling. This sort of inaccuracy is known as *sampling error*, which will be explored in more detail in the next chapter. Sampling errors are unavoidable, but the likely size of these errors can be estimated.

However, other sources of inaccuracy may exist that are not the result of sampling error. These are *nonsampling errors*, which could introduce systematic error into the estimation of population parameters. Nonsampling error leads to bias in the estimation process, and taking a larger sample of data will not necessarily reduce this bias. Every effort should be made to avoid nonsampling error, mainly because you can never know how large the bias might be.

Nonsampling errors can arise for many reasons, for example:

- Nonrandom sampling methods are used. Judgement introduces subjectivity, which can cause bias.

- The nonresponse rate is high. As mentioned earlier, this is a form of self-selection (or nonselection), and it is possible that those who choose to respond have a different opinion from those who do not. It is often better to have a smaller sample size and to use the money saved to increase the response rate.
- The target population is not clear, no adequate sampling frame exists, or the list is badly out of date.
- In a survey, some questions are ambiguous, misleading, or biased, or the interviewer is biased.

6.8 Chapter Summary

This chapter has been concerned with the basic concepts of sampling and the different methods that can be used to take a sample from a population. Specifically:

- The difference between a sample and a census, and their relative advantages and disadvantages
- The concept of a population parameter and a sample statistic
- The difference between random and nonrandom sampling
- A comparison of the accuracy of random and judgement sampling
- Methods of random sampling:
  - simple random
  - stratified
  - systematic
  - cluster
- Methods of nonrandom sampling:
  - quota
  - judgement
  - self-selected
- The importance of having a sampling frame
- The accuracy of sample estimates

6.9 Data for the Wood Experiment

The weights of the population of 100 blocks of wood, or equivalently the areas of the shapes in Figure 6.3, are given in Figure 6.9.

6.10 Exercises

1. Considering the four methods of random sampling described in Section 6.5, suggest an appropriate sampling scheme for the following two situations.
   a. How would you take a random sample of 1000 people from an electoral roll in order to estimate the support for different political parties?
   b. The University of Loughborough wishes to interview a random sample of its students to find out why they chose to study at the university. How might such a sample be chosen?

Block Number	Weight (g)	Block Number	Weight (g)	Block Number	Weight (g)	Block Number	Weight (g)	Block Number	Weight (g)
1	45	21	40	41	45	61	85	81	30
2	105	22	30	42	35	62	25	82	30
3	30	23	10	43	65	63	85	83	15
4	30	24	35	44	40	64	10	84	30
5	50	25	45	45	15	65	55	85	10
6	15	26	50	46	40	66	10	86	45
7	25	27	30	47	40	67	30	87	10
8	90	28	30	48	25	68	60	88	45
9	75	29	35	49	40	69	25	89	10
10	50	30	35	50	45	70	70	90	60
11	40	31	10	51	45	71	50	91	15
12	45	32	30	52	15	72	25	92	85
13	50	33	70	53	35	73	15	93	50
14	75	34	40	54	45	74	10	94	5
15	30	35	50	55	25	75	35	95	20
16	50	36	35	56	40	76	40	96	55
17	50	37	60	57	45	77	30	97	10
18	55	38	10	58	25	78	40	98	30
19	45	39	10	59	70	79	15	99	55
20	35	40	55	60	65	80	70	100	45

Figure 6.9   Weights of the population of wood blocks.

2. Suppose that British Airways is interested in carrying out a market research survey to find out its customers' opinions of its catering services.
   a. How should this customer survey be carried out?
   b. How should its sample its customers?
   c. What type of questions should be asked (e.g., open ended, yes/no types, using 5-point scales) and why?

3. On September 17, 1997, a referendum was conducted in Wales on the question of parliamentary devolution. People on the electoral register in Wales were eligible to vote in this referendum. They were asked to agree or disagree with the following proposition:

   I agree there should be a Welsh assembly.

   The result, from a turnout of 50.1%, was as follows.

Yes	559,419	50.3%
No	552,698	49.7%
Majority in favor	6,721	0.6%

   Comment on the outcome of this referendum.

4. A total of 237 students from a management statistics course handed in feedback forms at the end of the course. To analyse the open-ended questions, a systematic sample was taken. A sample of about 50 was the maximum that could be processed.
   a. Describe how to take a systematic sample from the population of forms.
   b. The forms are stored by tutorial group, with about eight students per group. Do you think this ordering of the forms would make the sampling method appropriate or inappropriate? Justify your answer.

5. Last year, the manager of North Island Airlines purchased health insurance for all of the company's employees. The company wants to find out the average weekly medical expenses for the employees during this year. There are 397 maintenance staff, 614 cabin staff, 162 senior pilots, and 27 management staff. A sample size of 100 is to be used.
   a. Which sampling method would you use to estimate the mean weekly medical cost per employee?
   b. Explain how you would allocate the sample between the four categories of employees.
   c. If it transpires that the medical costs for pilots are 40% more variable than for the other three groups, how would this alter your sample allocation?

6. State whether each boldfaced number in the following exercises is a parameter or a statistic.
   a. The National Statistics Office last month interviewed 25,000 members of the labour force, of whom **8.6%** were unemployed.
   b. A consignment of widgets has a mean weight of **15.4 g**. This is within the specifications for acceptance of the consignment by the purchaser. However, the purchaser falsely rejected the consignment because a randomly chosen sample of 50 widgets had a mean weight of **16.2 g**, which is outside the specified limits.
   c. A telephone sales company based in Swindon uses a device that dials residential phone numbers in that city at random. Of the first 200 numbers dialed, **19** are unlisted. This is not surprising, because **8%** of all Swindon phone numbers are unlisted.
   d. A market researcher investigating the reactions to a new slimming food conducts a comparative experiment with 20 randomly chosen subjects. A control group of 10 is fed a placebo, while the experimental group of 10 is fed the new product. After four weeks, the mean weight loss is **14 kg** for the control group and **25 kg** for the experimental group.

7. Jenny, the marketing manager of the telemarketing company EazySell, wants to determine the average number of telephone calls her operators make each month. There is little variation in the number of calls within a month but considerable variation from month to month. Jenny also suspects that they make more calls earlier in the week than later. Explain carefully what sampling method Jenny should use to accurately estimate the overall average number of calls made monthly.

8. In the week preceding a local election, the editor of the *Loughborough Echo* decides to take an opinion poll to try to predict the outcome of the election. Four possible methods are proposed for selecting a sample of voters:

     i. Interview every 20th person over 18 entering a local supermarket between 9:00 a.m. and 5:30 p.m. this coming Friday.

    ii. Select three streets at random from the Loughborough street directory and interview every person over 18 in each household. These interviews will be conducted over the next four evenings.

   iii. Select every tenth page from the local residential telephone directory and call the first number with a Loughborough prefix (01509) listed on each page. The calls are to be made over the next four evenings, and information will be collected from whomever answers the phone, provided that person is over 18.

   iv. Two interviewers will interview people in the Curzon cinema lobby this coming Saturday between 12 noon and 10 p.m. One interviewer will select only males, and the other only females. They will interview as many people over 18 as they can in the time available.

  a. Classify each of the proposed sampling schemes as either random or nonrandom.

  b. Classify each of the random schemes as simple random, stratified, systematic, or cluster.

  c. Classify each of the nonrandom schemes as quota, judgement, or self-selected.

  d. Give a possible source of bias associated with each sampling scheme (i.e., a reason why it may not produce data that are representative of the voting population).

9. The management of Menser and Sparks is seeking to promote the use of the store's loyalty card. The management has decided to survey a sample of existing cardholders to determine what factors influenced their decision to apply for a card and whether their shopping habits are affected by promotions to cardholders.

  a. Describe how you could take a systematic sample of cardholders.

  b. Describe how you could take a cluster sample of cardholders.

  c. If you chose to take a stratified sample of cardholders, indicate, with reasons, the three most important factors to take into account in designing your strata.

  d. Which of the three sampling methods would you recommend in this case? Give your reasons.

10. A major toy retailer in Leicester wants to attract new customers by modifying and enhancing the in-store facilities. As a part of the planning process, the management of the store decides that a survey should be undertaken to ascertain what customers would like to see to improve the store's attractiveness. It is decided to conduct this survey using personal interviews, but management is unsure what process to use.

  a. What population should the store attempt to survey? Explain why it would be inappropriate to attempt a census of this population.

  b. Give two reasons why the following sampling methods do not give a random sample of the population concerned.

    i. Select every 50th person passing the store on a particular Wednesday.

    ii. Select every child in seventh grade at the local school.

   iii. Select a quota sample of 60% women and 40% men entering a similar toyshop in a city 15 miles away.

    c. What sampling method would you suggest? Discuss in detail how you would select the sample that you recommend.

11. The manager of a large supermarket has become aware that the store is receiving more damaged and dented cans than previously. Any badly damaged cans tend to get left on the shelves and ultimately have to be sold cheaply or returned to the supplier. To find out how extensive the problem is, he decides to sample the incoming deliveries of canned products over the next two weeks.

    a. Discuss some of the factors you would consider in stratifying this situation.

    b. Describe in detail how you would design a sampling scheme to take account of the various factors you have identified in (a).

    c. What would be your "measure" of the scale of the problem? How would you display the results of your investigation to senior management?

12. Tesda, a large retail group, is investigating the feasibility of staying open all night in selected stores. The manager of a possible all-night store has been told to conduct a survey of her customers to find out their attitude to all-night shopping. The store manager has some information about her customer population, based on holders of loyalty cards, from which she determines the percentage breakdown by age and gender shown in Figure 6.10. In a typical week there are about 6000 customers. The store manager has a limited budget to conduct the survey, and head office has asked for a report on the results of the survey within 3 weeks. In light of this, she decides that a sample of 200 customers is possible within the time and budget available.

    a. Would a simple random sample of customers be feasible? Why or why not?

    b. Indicate how a quota sample might be chosen from the customer population.

    c. How could a cluster sample of customers be chosen? What factors should be taken into account when considering cluster sampling?

    d. How could a systematic sample of customers be chosen? What factors should be taken into account when considering systematic sampling?

    e. Considering the four possibilities in (a) to (d), indicate which method you would use, giving reasons for your choice.

13. The Code of Social Responsibility was sent to every householder in New Zealand. The government expected 10% of the population to fill in and return the enclosed questionnaire.

Sex	Age group				
	Under 25	25 – 34	35 – 44	45 – 59	60 and over
Male	2	4	8	7	3
Female	6	15	25	20	10

Figure 6.10   Percentage age and sex distribution of loyalty card holders.

a. Do you agree or disagree with the statement that the data collected will be worthless? Explain your answer.

b. If the 10% of respondents are regarded as a sample from the population, which type of sampling scheme has been used?

c. As an alternative, the government could have taken a random sample of households. Discuss briefly how you might take such a sample, and comment on how the sample size should be chosen.

14. A student from the management school at the University of Waikato is studying the price of a brand of chocolate bar. He makes a list of all the food shops and vending machines within a 3 km radius of the university campus and their locations. He then samples at the 10 locations nearest his home.

a. What type of sampling is taking place?

b. In what way could the sampling scheme give misleading results?

c. How would you sample more effectively from the list of food shops and vending machines?

Chapter 7

Estimation

7.1 Sampling Error

We have seen in earlier chapters that data are usually collected in order to gain *information* about, and better understand, some particular population or process. In some circumstances it may be appropriate to attempt to collect data about every member of the population (a *census*), but this is often impractical because many of the populations we wish to study are impossibly large, or the sampling method is destructive. It is therefore more usual to collect information from only a sample of the population, as described in the previous chapter. In practice, samples of data may be collected in many ways, both random and nonrandom, but however the sample is taken, it provides data for only a portion of the entire population. So it is unlikely that a sample will give precise information about the population. Usually error will arise from using a sample to estimate population characteristics. With a random sample, this error is known as the *sampling error*. With a nonrandom sample, however, estimates may also be *biased,* in that a systematic error may arise because judgement has been used in the sampling process.

Whenever we rely on sample information it is important to have an appreciation of how large the sampling error might be. Without an understanding of the sampling error, we are likely to misuse the information and possibly make wrong decisions.

⌂ *Thorcam Domestic Appliances*

One of the key components in Thorcam washing machines is the timer that controls the washing cycle. In recent months the number of customer claims has been higher than usual for timers failing within the 1-year warranty period. Such claims result in an expensive replacement or repair.

To assess the life of timers under normal operating conditions, a sample of 25 is tested by being run until failure. The number of hours of continuous operation

Timer	Life	Timer	Life	Timer	Life	Timer	Life	Timer	Life
1	470	6	842	11	684	16	694	21	658
2	702	7	1020	12	696	17	516	22	898
3	814	8	968	13	580	18	400	23	670
4	98	9	950	14	360	19	912	24	654
5	548	10	988	15	594	20	900	25	534

Figure 7.1   Life of Thorcam washing machine timers (hours).

until failure is recorded in Figure 7.1. The average (mean) life is 686 hours, and the standard deviation is 224.4 hours.

> **Q1.**   *What can you conclude about the mean life of the timers?*

📈 *Wood Experiment*

In the wood experiment in Chapter 6, students are asked to select two samples of 10 blocks of wood, the first using their judgement and the second using a random sampling approach. The mean block weights from the random samples of a group of 171 students are shown as a histogram in Figure 7.2.

In practice, if we had to estimate the mean weight of the wood population, we would take just one sample of blocks rather than repeat the experiment 171 times.

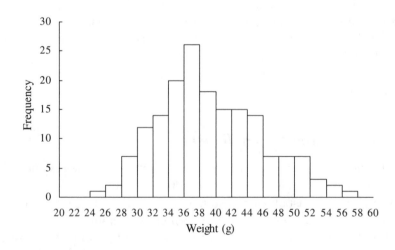

Figure 7.2   Wood experiment — random sample means.

However, the distribution of results in Figure 7.2 tells us what range of mean values we are *likely* to get from a random sample of 10 blocks of wood.

> **Q2.** *From the data in Figure 7.2, what are the minimum and maximum values we could realistically get for the mean weight of a random sample of 10 blocks of wood?*
>
> **Q3.** *Would it be the same if we took a sample of 40 blocks of wood?*

In general, we are concerned with the accuracy of sample information when estimating some population *parameter*. In other words, how large is the sampling error? In the two preceding examples, we have interval data and our concern is with the population *mean*, that is, the mean life of timers or the mean weight of blocks of wood. Later we shall consider binary data where we wish to estimate the *proportion* of cases in which a particular attribute occurs. Although we are usually interested in estimating a mean or a proportion, we might be concerned with other population parameters, for instance:

- The population median, for example, the median travelling time to the university
- The population standard deviation, for example, of the volume of coke in a can
- The population mean for a value of another variable, for example, the average salary of accountants at age 40

In this chapter we will concentrate on sampling errors in estimates of a population mean and a population proportion.

7.2 Estimating a Mean

We first consider the problem of estimating the mean of a population from a random sample of interval data. Sampling error is always involved in any estimate of a population mean, but how do we quantify it? For example, how accurate is the mean life of the sample of 25 timers in Figure 7.1 as an estimate of the mean life of all timers? To answer this question we need to know how much variation will be found in the distribution of a (large) number of such sample mean values.

This is exactly what we did in the wood experiment. We took a number of random samples of size 10 from the population of blocks of wood and calculated the mean of each sample. We then constructed a histogram of these means, an example of which is shown in Figure 7.2. If we have a large enough number

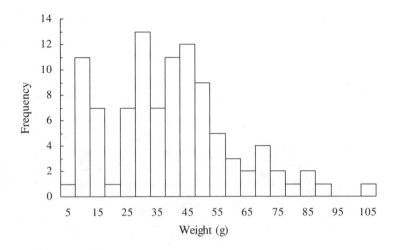

Figure 7.3   Distribution of the weights of 100 blocks.

of samples, we can approximate this plot by a smooth distribution curve. This distribution is called the *sampling distribution of the mean*.

⚡ *Sampling Distribution of Mean Weights*

How does the sampling distribution of the mean compare with the distribution of the population from which the samples were taken? In the wood experiment, we know that the population consists of 100 blocks with weights ranging from 5 g to 105 g, as shown by the histogram in Figure 7.3.

Q4. How does the average, spread, and shape of the sampling distribution of the mean (Figure 7.2) compare with the population distribution (Figure 7.3)?

Q5. What do you think would happen to the sampling distribution of the mean in Figure 7.2 if we were to take smaller samples (say, 5) or larger samples (say, 20)?

Using random sampling methods, we can generate sampling distributions corresponding to different sample sizes (n). To give a direct comparison with Figure 7.2, we have taken 171 independent random samples of sizes 5 and 20 from the population of 100 blocks shown in Figure 7.3. The mean weight of the items in each sample was calculated, and the 171 sample means are shown in the histograms in Figure 7.4, together with the histogram for samples of size 10.

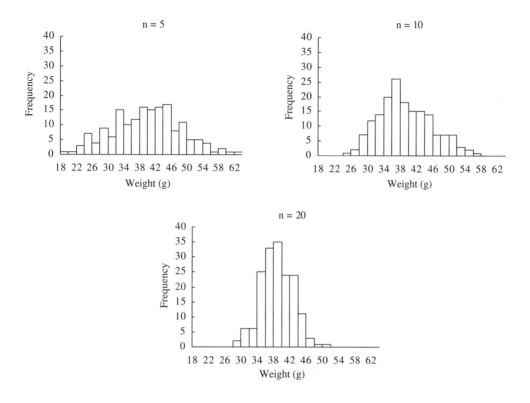

Figure 7.4   Sampling distributions of the arithmetic mean for $n = 5, 10,$ and 20.

Q6. What can you say about the location, spread, and shape of these distributions?

✎ *Properties of the Sample Mean*

Figure 7.4 shows how the sample mean $(\bar{x})$ varies for different sample sizes n. These distributions illustrate the following properties of the sampling distribution of the mean:

- The sampling distribution of the mean tends to be *bell shaped* and approximately normal, irrespective of the shape of the distribution from which the sample of data is taken. The normality of the distribution improves as the population distribution itself becomes more normal and as the sample size increases. In practice, it is safe to assume that the distribution of $\bar{x}$ will be normal if
  - the sample size (n) is greater than 30 or
  - the sample size (n) is greater than 10 and the distribution of the observations is reasonably symmetrical or
  - the distribution of the observations is close to normal, for any value of n.

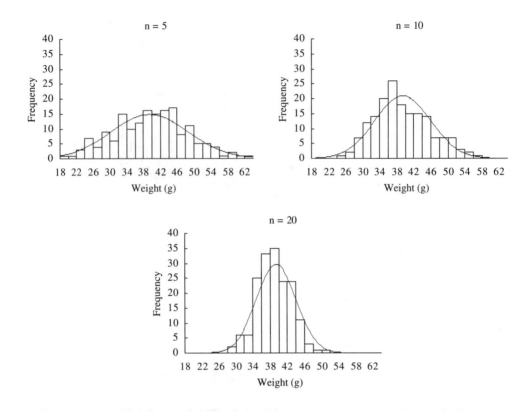

Figure 7.5 Sampling distributions of the mean with normal distributions.

- The *mean* of the sampling distribution is the same as the population mean μ, for any value of n.
- The *standard deviation* of the sampling distribution is $\sigma/\sqrt{n}$, where σ is the standard deviation of the population. The standard deviation of the sampling distribution is usually called the *standard error* of the mean.

For the population of weights in Figure 7.3, the mean of all 100 items is actually 39.4 and the standard deviation is 20.63. So if samples of size 5 are taken from this population, the sample means should average 39.4, with a standard deviation of $20.63/\sqrt{5} = 9.23$. Alternatively, with samples of size 10, the standard deviation should be $20.63/\sqrt{10} = 6.52$, and for samples of size 20, it should be $20.63/\sqrt{20} = 4.61$.

Normal distributions with a mean of 39.4 and with the appropriate standard deviation are shown in Figure 7.5 superimposed on the relevant histogram from Figure 7.4. It can be seen that these normal models represent the data well, despite that fact that the sample sizes (5, 10, and 20) are not particularly large.

Q7. *Remembering that the mean of the population is actually 39.4, how likely is it that the mean of a random sample of 10 items would be less than 25?*

7.3 Margin of Error for a Mean

To understand the concept of sampling error, it is important to realise that a given sample result is only *indicative* of the overall situation and will not necessarily reflect exactly the values for the population as a whole. We cannot conclude that the mean weight of all the blocks of wood is 37.5 g just because a random sample of 10 pieces of wood gave a mean of 37.5 g. If we took another sample of 10 blocks, we might get a mean of 42.5 g. A further sample of 10 blocks might yield a mean of 39.0 g. In fact, Figure 7.2 shows that when we took 171 random samples there was a great deal of variability in our sample means.

Because we do not expect the sample mean to be exactly equal to the population mean μ, we need to provide information about the accuracy of our estimate. This is accomplished by stating a *margin of error* for our estimate of the population mean, which indicates the extent to which sample results can vary from the true figure.

In particular, the margin of error gives us:

- An interval of uncertainty within which the population mean *probably* lies.
- Some idea of how often the population mean would be expected to be within this interval in repeated sampling. We specify the strength of "probably" in terms of a confidence level.

From Appendix A.2, 95% of the values of a normal distribution fall within 1.96 standard deviations either side of the mean value. Hence, knowing that a sample mean is approximately normally distributed with a standard deviation of $\sigma/\sqrt{n}$ means that 95% of sample means will fall within a margin of error of

$$\pm 1.96 \times \frac{\sigma}{\sqrt{n}}$$

In other words, 95% of the time the difference between a sample mean and the population mean will be less than this margin of error. For example, given that the standard deviation of the weights of items in Figure 7.3 is 20.63, the 95% margin of error for the mean of a sample of 10 items is

$$\pm 1.96 \times \frac{20.63}{\sqrt{10}} = \pm 12.8$$

Q8. *Explain in simple terms what this margin of error means.*
Q9. *Would you expect the 95% margin of error for a sample of five items to be more or less than 12.8? Verify your answer by performing the calculation.*

Of course, we do not have to use a percentage of 95%. In some cases we may want to be more sure of our assertions and so choose (say) a 99% confidence level or possibly one even higher than that. Alternatively, we might be content to accept a lower level of confidence in our assertions, although it is unusual for the confidence level to be less than 90%. In general, we can determine a margin of error for any confidence level by using the relevant percentage point (z-value) of the standard normal distribution from Appendix A.2 as:

$$\pm z \frac{\sigma}{\sqrt{n}}$$

This margin of error is appropriate when:

- A random sample of observations has been taken.
- The population standard deviation (σ) is known.
- The sampling distribution of the mean is approximately *normal*, that is, the population being sampled is itself approximately normal or the sample size is sufficiently large (i.e., about 30 or more).

Q10. How does the margin of error alter if the confidence level increases or decreases?

Q11. For 90% confidence, what value do we use in place of 1.96? What value for 99% confidence?

Q12. Could we be absolutely certain by using a confidence level of 100%? Explain why or why not.

Thorcam Domestic Appliances

Recall the problem at the beginning of this chapter concerning the life of Thorcam washing machine timers. A random sample of 25 timers, when tested to failure, gave an average life of 686 hours with a standard deviation of 224.4

hours. If we can assume that the standard deviation of the life of all timers is about 220 hours, then the 95% margin of error for the sample mean is

$$\pm 1.96 \times \frac{220}{\sqrt{25}} = \pm 86.2$$

i.e., ±86 hours

Q13. Explain in plain language what this margin of error means.
Q14. Calculate the 99% margin of error for the mean life of Thorcam timers.

✎ Confidence Interval

When specifying a margin of error, we are really saying that a range, or interval, exists around the sample result within which the population parameter *probably* lies. Either this interval of values contains the population parameter or it does not. We have a certain level of belief, or *confidence*, that the population parameter falls within this interval. For this reason, the interval is called a *confidence interval*.

A 95% confidence interval for the population mean life of Thorcam timers is 686 ± 86, i.e., between 600 and 772 hours.

7.4 Interpreting the Margin of Error

Suppose that in a large class each student calculated the mean weight of wood for a random sample of 10 blocks of wood, giving a distribution of means similar to that in Figure 7.2. The 95% margin of error of the mean is ±12.8. This means we would expect, for about 95% of the students, the difference between their sample mean and the population mean of 39.4 g to be between −12.8 and +12.8 g. In other words, if the margin of error was applied to their mean value, each student would have a confidence interval, and we would expect about 95% of these intervals to include the population mean. Of course, as we said earlier, for any given student that interval would either include the population mean or not.

A simulated exercise, one that reinforces what can be seen in the wood experiment, is given in the spreadsheet demonstration *ConfInt.xls* on the CD-ROM. In it a random sample is taken from a normal population with mean 20 and standard deviation 5. Figure 7.6 shows the spreadsheet after a single

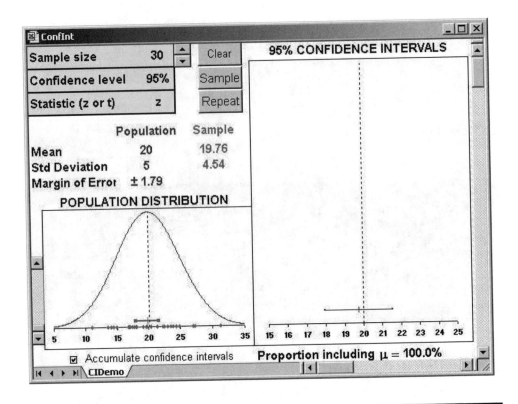

Figure 7.6   Output from the *ConfInt* demo spreadsheet.

sample of size 30 has been taken. The sample is presented on the horizontal axis of the normal distribution, and the sample mean is shown to be 19.76. The 95% margin of error of ±1.79 is calculated using $z = 1.96$. The confidence interval 19.76 ± 1.79 appears in the population distribution and, if the *Accumulate* box below the population distribution is ticked, also in the 95% confidence interval panel on the right-hand side. We see that the interval covers the population mean of 20.

Repeatedly clicking the *Sample* button will give further samples of size 30, with margins of error calculated and confidence intervals displayed. In Figure 7.7 we see the results obtained from taking 20 separate samples. In this case only the seventh sample has produced a confidence interval that does not include the population mean, $\mu = 20$, so the proportion including μ is 19 out of 20, or 95%.

Clicking *Clear* will clear all calculations. Clicking *Repeat* will generate 100 samples and display the confidence intervals. Now we would expect to get about five intervals that do not include the population mean. Both the sample size and the confidence level can be changed in the spreadsheet. This allows you to explore what happens if the sample size is changed or if a different confidence level is chosen. Notice the effect this will have on the size of the margin of error and on the proportion of confidence intervals that include the population mean. The *z*-value can also be replaced by an alternative *t*-value, which is considered in Section 7.6.

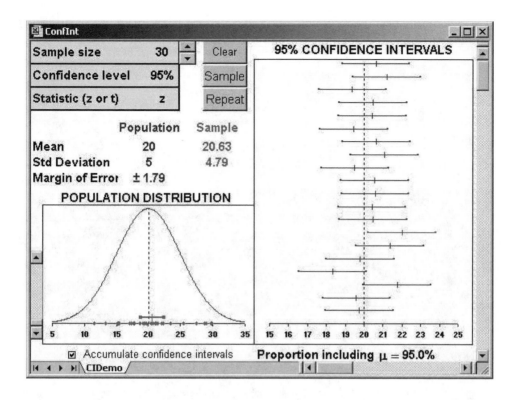

Figure 7.7   A sample of 20 confidence intervals using z.

7.5   Sample Size for Means

The margin of error of the mean is inversely proportional to the sample size n. That is, as the sample size increases, the margin of error decreases.

Q15. *Is this what you would expect? Why or why not?*

Thus, by rearranging the formula for the margin of error, we can determine the sample size required for a specified margin of error:

$$n \geq \left(\frac{z \times \sigma}{\text{margin of error}} \right)^2$$

where z is taken from Appendix A.2 for the desired confidence level.

For example, how many timers need to be tested to estimate the mean life of all timers to within 40 hours? If we assume a 95% margin of error, then $z = 1.96$, so

$$n \geq \left(\frac{1.96 \times 220}{40} \right)^2 = 116.2$$

i.e., 117 timers.

Q16. *Suppose we want to be 95% confident that the sample estimate will be within 20 hours of the population mean. What sample size would be required?*

Q17. *What general conclusions can you draw about the relationship between the margin of error and the sample size required?*

Q18. *In particular, to halve the margin of error, we must increase the sample size by what factor?*

In practice it is very hard to get an accurate estimate of the standard deviation σ, so the sample size we calculate is usually at best only approximate. But it is better to have a general idea of how big a sample you might need to take, and the costs associated with it, before the event. It might be appropriate in some cases to take a small sample to get a better estimate of the population standard deviation and to use this to determine the sample size more accurately. The initial sample can then be "topped-up" to the required number. Of course, other factors influence sample size, such as cost and available resources. However, experience suggests that people tend to overestimate the sample size they require. Unnecessary money and resources can be wasted collecting data from very large samples.

7.6 Unknown Population Standard Deviation

In practice, we seldom know the population standard deviation σ. In this case we make two changes to the margin of error:

1. As you might expect, we now use the sample standard deviation (s) in the formula to replace σ.

2. As a result, because *s* varies from sample to sample, the margin of error will also be subject to uncertainty. To compensate for this, in the formula for the margin of error we replace the *z*-value we looked up from the standard normal distribution (Appendix A.2) with a value from Student's *t*-distribution.

Student's t-distribution, or simply the *t-distribution*, is another commonly used statistical distribution. It is also a symmetric distribution and has one parameter, called the *degrees of freedom*. The concept of degrees of freedom was discussed in Section 4.4 in relation to the standard deviation, where the number of degrees of freedom was $n - 1$. The number of degrees of freedom in the calculation of *s* is simply an indicator of how good *s* will be as an estimator of σ. The bigger the sample, the more degrees of freedom it will have and the better *s* will be as an estimate of σ. It is this quantity that determines the value of *t* we use to replace *z* in our formula for the margin of error. For a particular confidence level, *t* will always be greater than *z*; but as the degrees of freedom get larger, *t* gets closer and closer to *z*.

The margin of error for the population mean μ, based on a random sample of size *n*, is

$$\pm t_{n-1} \frac{s}{\sqrt{n}}$$

where t_{n-1} is an appropriate percentage point of the *t*-distribution with $n - 1$ degrees of freedom.

This margin of error is valid provided that:

- The *n* observations are a random sample from a *normal* population.
- The unknown population standard deviation is estimated by the sample standard deviation *s*.

A table of selected percentage points of the *t*-distribution is in Appendix A.3. The table gives the value of *t* such that a specified percentage (*P*) of the distribution lies between −*t* and +*t*, for *P* equal to 90, 95, 98, or 99 and for different numbers of degrees of freedom (*df*). In other words, the appropriate row of Appendix A.3, instead of the values in Appendix A.2, gives us the required percentage points.

♨ Excel Demonstration

The *ConfInt.xls* spreadsheet can also be used to simulate what happens when a margin of error for the mean is obtained using the sample standard deviation. Figure 7.8 shows the results of taking 20 samples, each of size 30, and plotting the confidence intervals. Note that the *t* statistic has now been chosen in the spreadsheet. The final sample shown has mean 19.83, a standard deviation of 4.54, and a margin of error of ±1.69. The resulting confidence interval includes the population mean μ. Of the 20 samples, two give confidence intervals that

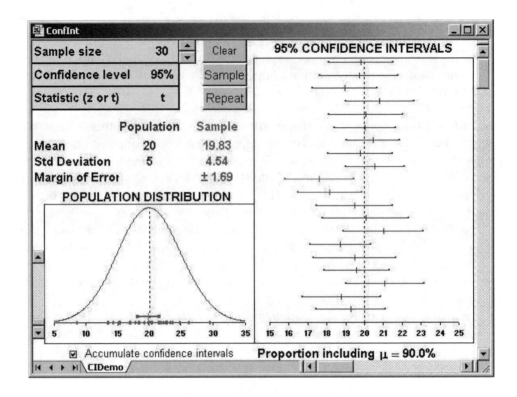

Figure 7.8   A sample of 20 confidence intervals using *t*.

do not include μ, so the proportion including μ is 90%. We would expect this proportion to be about 95% if we continued sampling. Again you can use the spreadsheet to investigate what happens if you take more samples (using the *Repeat* button, for example) or change the sample size and confidence level.

One obvious difference between the results in Figures 7.7 and 7.8 is the size of the margins of error (or widths of the confidence intervals). In Figure 7.7 all margins of error are the same, namely, ±1.79. In Figure 7.8 the margins of error vary in magnitude. This is because they involve the sample standard deviation, which changes from sample to sample. In the plot of the confidence intervals we can see this reflected as intervals of different widths. Even so, we would still expect about 95% of the intervals to include the population mean.

Thorcam Domestic Appliances

In our previous calculation to determine the margin of error for the life of Thorcam washing machine timers, we made the (perhaps rather bold) assumption that the standard deviation of the population of timers was about 220 hours. In reality, we do not know what the population standard deviation is, and it may *not* be 220. However, using the *t*-distribution, we can now work out an exact margin of error using the *sample* standard deviation $s = 224.4$, with $25 - 1 = 24$ degrees of freedom.

For a 95% confidence level, $t_{24} = 2.064$, giving a margin of error

$$\pm 2.064 \times \frac{224.4}{\sqrt{25}} = \pm 92.6 \text{ hours}$$

With 95% probability, the sample mean (686 hours) should be accurate to within 92.6 hours. The mean life of all timers should, therefore, be within the range 686 − 92.6 = 593.4 hours to 686 + 92.6 = 778.6 hours.

From our analysis of the sample of data in Figure 7.1 it is apparent that considerable uncertainty exists in our estimate of the average life of Thorcam washing machine timers. Our sample estimate of 686 hours could be inaccurate by as much as about 100 hours. Two factors mainly contribute to this uncertainty: the relatively small sample size and the large amount of variation in the lives of timers. If Thorcam wishes to have more control over the early failure of timers and the consequential warranty costs, it must clearly address the issue of variation in timer life. In Chapters 11–13 we show how timer life could be monitored using techniques of statistical process control and how the apparently large variation might be reduced.

7.7 Finite Populations

So far we have implicitly assumed that the population being sampled is sufficiently large that any sample we take constitutes a tiny, insignificant fraction of the whole population. What if this was not the case? Is there a difference between a sample of 50 from a population of millions and a sample of 50 from a population of 100?

> *Q19. What do you think? Does the population size affect the margin of error? If so, how?*

To see that population size does matter, you only have to think of the extreme case when we sample the entire population. For example, if we take a sample of 100 items from the wood population in Figure 7.3, we have observed every item in the population, and so any sample statistics we calculate (such as the mean) will be totally accurate, with no error. Whenever we take a sample from a finite population, where the sample constitutes a nonnegligible fraction of the population, we must amend our margin of error formula by multiplying it by the *finite population correction factor,* given by

$$\sqrt{1 - \frac{n}{N}}$$

where N is the size of the population. The correction factor is the square root of the fraction of the population that is *not* sampled.

📖 Wood Experiment

Suppose we take a random sample of 10 items from the population in Figure 7.3, obtaining the following values:

10, 35, 40, 20, 75, 60, 40, 55, 10, 85

> **Q20.** *Assuming we know that the standard deviation of the population is 20.63, calculate a revised 95% margin of error incorporating the finite population correction factor.*
>
> **Q21.** *Assuming we do not know what the population standard deviation is, determine the new margin of error using the sample standard deviation 25.8.*

7.8   What Affects the Margin of Error?

The margin of error is dependent generally on three key factors of the sampling situation in question:

1. The amount of variation in the population being sampled. The more variable the population, the greater will be the margin of error in any sample estimate.
2. The size of the sample. The larger the sample, the smaller the margin of error. The main point to note here is that sampling error is related to the *square root* of the sample size, and so to halve the error, we must quadruple the sample size.
3. The chosen level of confidence reflected in the particular percentage point (z or t) used. A greater confidence level (more confidence in the conclusion) will increase the resulting margin of error.

As described in the previous section, where the population being sampled is small, then the fraction of the population that is sampled also has an effect, but this is seldom important because populations are usually large, if not infinite.

The first factor just given may be difficult to control, because it is inherent in the population or process being studied. So to reduce the margin of error, we must either increase the sample size or sacrifice some confidence in the results. If the sample size is also fixed or predetermined, then decreasing the level of confidence is the only way to reduce the margin of error.

7.9 Estimating a Proportion

So far we have considered the problem of estimating the population mean from a random sample of interval data. Consider now the following three examples.

📖 *Value Added Tax (VAT)*

The Tax Department sampled 350 VAT returns and found that 43 had major errors in them.

> **Q22.** *What can you say about the proportion of all VAT returns with major errors?*
> **Q23.** *What other information would you like to have?*

📖 *Union Membership*

A random sample of 50 small businesses in a city of 200,000 inhabitants found that only 12 businesses had workers who were members of a trade union.

> **Q24.** *What conclusions can you draw from this sample about trade union membership in small businesses in that city?*
> **Q25.** *What conclusions can you draw about trade union membership nationally?*

📖 *An Opinion Poll*

The following statement is from a local newspaper just prior to a national election.

> In an opinion poll of 1000 electors, one of the political parties has a support of only 2%. This is below the margin of error (of 3%).

> **Q26.** *What you think this statement implies? Does it make sense?*

In these three examples we are concerned with estimating the true *proportion* of VAT returns with errors or small businesses with trade union members or people voting for a political party. To calculate the margin of error for a population proportion, we could, in principle, take a succession of random samples of a given size and see how much variation exists in the sample proportions we get. But this is not necessary because statistical theory can again be used to infer what this variation would be. In Section 4.12 we saw that a sample proportion is approximately normally distributed, provided the sample is reasonably large, with a standard deviation given by

$$\sqrt{\frac{p(1-p)}{n}}$$

where p is the expected proportion, that is, the true proportion for the entire population. Hence, the margin of error of a proportion is given by z times the standard deviation:

$$\pm z \sqrt{\frac{p(1-p)}{n}}$$

Thus, a specified percentage of the time the difference between a sample proportion and the true proportion will be less than this margin of error. Appendix A.2 is again used to obtain the z-value for the chosen level of confidence.

📇 An Opinion Poll

Think again about the quote from the newspaper regarding the political party having 2% support in an opinion poll of 1000 electors. Given an estimated proportion of 2%, $p = 0.02$, so the 95% margin of error is

$$\pm 1.96 \sqrt{\frac{0.02 \times (1-0.02)}{1000}} = 0.009$$

i.e., ±0.9%.

Q27. *Where do you think the figure of 3% has come from? Could it be based on a 99% level of confidence?*

7.10   Sample Size for Proportions

By rearranging the formula for the margin of error of a proportion, we can determine the sample size required for a specified margin of error:

$$n \geq \left(\frac{z \times \sqrt{p(1-p)}}{\text{margin of error}} \right)^2$$

where z is taken from Appendix A.2 for the desired confidence level.

Inevitably p will be an unknown quantity, for we are determining the sample size required to estimate p with a desired margin of error. When p is completely unknown, you should assume the worst-case scenario and use $p = 0.5$ because, as we saw in Figure 4.26, this is where the standard deviation, and hence the margin of error, is greatest. If p can be roughly estimated within a particular range, you should use the value of p nearest to 0.5. For example, if we can be fairly sure that p is somewhere between 0.1 and 0.3, then the value $p = 0.3$ should be used because this will lead to the largest value of n.

Note that for a 95% confidence level, $z \approx 2$, so $z \times \sqrt{p(1-p)} = 1$ when $p = 0.5$. This gives the simple result

$$n \geq \frac{1}{(\text{margin of error})^2}$$

for 95% confidence in the worst-case situation.

> **Q28.** *In a political opinion poll a 95% margin of error of ±3% is required. Show that a sample of about 1000 electors is needed to estimate the percentage vote for any party.*
>
> **Q29.** *What would be the 95% margin of error, using a sample size of 1000 electors, for a political party with a level of support of about 5%?*

7.11   Tests of Hypotheses

We collect data to find out about a population or process. In the Thorcam Appliances example, we wish to know the long term average life of timers. In other words, how long can we *expect* any timer to last until breakdown? We can conduct an experiment or test and use the results from this experiment to estimate the population mean with an appropriate margin of error.

Sometimes, however, we have an idea of what the population mean should be, based perhaps on some theory, experience with similar processes, or hunch. Then we might be interested in seeing whether the results of our experiment are consistent with this idea. For example, as a result of the increased number of warranty claims, we may have reason to question whether the performance of Thorcam timers has changed from the previously established level. This could be the result of a design change, changes to the manufacturing process, or a different type of washing machine in which the timers are now installed.

Thorcam Domestic Appliances

Thorcam claims that the timers used in its washing machines have a design life of 750 hours.

> **Q30.** *Do you think the results given in Figure 7.1 cast doubt on the claim of the supplier? Why?*

What we need to do is to carry out a *test of significance*, or *test of hypothesis*. Our hypothesis in the Thorcam example is that the population mean is 750. If we calculate a confidence interval and find that our hypothesised mean (750) does not fall within this interval, then the evidence suggests that our hypothesis is unreasonable or, equivalently, that a *statistically significant* difference exists between the sample mean and the hypothesised population mean. The value of 750 actually falls comfortably within the confidence interval calculated earlier, so there is no firm evidence to reject a mean value of 750. Of course the problem might be that we have not taken a large enough sample to detect a difference, and we might decide to collect some more data to see if this is the case.

In this particular case, though, Thorcam might be concerned by what appears to be excessive variability in the lifetimes of the timers. As said in Section 7.6, it should perhaps give attention to studying its processes, with a view to reducing this variability.

Red Beads Experiment

The red beads experiment involved 400 red beads and 1200 white beads. That is, one-quarter of the beads in the raw material are red. Since the paddle holds 50 beads, we might expect one-quarter of these 50 to be red, that is, 12.5 red beads on average, or a true proportion of 12.5/50 = 0.25. Are the results you obtained in Figure 1.5 from a sample of 25 consistent with this theory? Our hypothesis is that the true proportion is $p = 0.25$, leading to a 95% margin of error of

$$\pm 1.96 \sqrt{\frac{0.25(1 - 0.25)}{25 \times 50}} = \pm 0.024$$

If the results of your experiment had an overall proportion of red beads in the range 0.25 ± 0.024, that is, between 0.226 and 0.274 or an average number of red beads between 11.3 and 13.7, then you would have no reason to reject the hypothesis. In Figure 1.9 we have 425 samples giving a 95% margin of error of ± 0.006, so if the average number of red beads is less than 12.2 we would reject the hypothesis. It is clear from Figure 1.9 that this is the case, with the mean number of red beads from the 425 samples being 10.19. The reasons why the average is not 12.5 were discussed in Section 1.6.

✥ Margin of Error and Tests of Hypotheses

The concepts involved in testing hypotheses are similar to those behind the margin of error. We shall not develop them further here. In business the problem of estimation is far more important, or should be, than testing some hypothesis. The primary purpose of collecting data should be to find out about the population or process under study, not to test whether some theory or other is valid.

However, hypothesis tests can sometimes be a useful decision tool. But you should bear in mind the following general points about margins of error and hypothesis tests:

- A margin of error:
  - gives *quantitative* information about the range of uncertainty in some population or process parameter.
  - demonstrates the *accuracy* of some estimate by the size of the margin of error.
- Hypothesis testing:
  - gives *qualitative* information about the correctness of some belief regarding a population or process parameter, but it
  - cannot totally confirm or deny such a belief, and it
  - can be fallible when there is a great deal of variation in the process, as you may be misled into thinking that the data are consistent with the hypothesis when really there is too much variation to be able to tell.

Brian Joiner, an eminent statistician and management consultant, says:

> I would say never test a hypothesis unless you're desperate and can't think of anything else to do. When you're testing a hypothesis, just knowing whether that hypothesis is to accept or reject is not a very informative statement. It doesn't give you useful information as a basis

for action.... It's a big mistake for statisticians to be teaching those kinds of things. It's doing serious damage. (*Quality Progress*, 1988, pp. 31–32.)

Hence our advice is to use estimation with a margin of error.

7.12   Case Study: Advertising Awareness

An advertising agency has been commissioned to develop a TV advertising campaign for a major bank. The campaign involves a new 30-second advertisement appearing regularly between 7:30 and 9:00 p.m. three times a week over a six-week period. To test the impact of the advertisement, the agency has decided to run a trial screening on national TV at around 8:00 p.m. next Wednesday and Saturday evenings. After each screening, the agency surveys its viewers' panel of 386 randomly chosen people to determine whether they were watching TV at the time and if so whether they were aware of having seen the advertisement by being able to recall the name of the bank in question. The results obtained are shown in Figure 7.9.

The data in Figure 7.9 are attribute data, with each member of the panel being aware of the advertisement or not. This can be summarised as a proportion of those watching TV who were aware of the advertisement on each day. For example, 121 out of 216 panel members watching TV were aware of the Wednesday advertisement, a proportion of 0.56 (i.e., 56%). This proportion rose to 0.68 (i.e., 68%) following the Saturday screening. Combining the two days results gives an overall awareness of 62.2%. But is this valid?

The 95% margin of error for the Wednesday advertisement is

$$1.96\sqrt{\frac{0.56(1-0.56)}{216}} = 0.066$$

i.e., 6.6%. Therefore, we can be 95% confident that between 49.4% and 62.6% of all viewers are aware of having seen the Wednesday advertisement. A similar calculation shows that the margin of error for the Saturday proportion is 0.061, meaning that the proportion of viewers aware of this advertisement is between 61.9% and 74.1%.

So do we have clear evidence that the advertisement has created significantly greater awareness on Saturday than on Wednesday? The two confidence intervals overlap slightly, so it is possible that the awareness for both advertisements is about 62%. But is this reasonable? For this to be true, the Wednesday advertisement would need to be at the top end of its feasible range and the Saturday awareness to be somewhere near its lowest conceivable value. In other words, two relatively unlikely events both need to occur.

In this case, if we present the data from Figure 7.9 in the form of a two-way table, as in Figure 7.10, we can use the chi-square test (described in Chapter 5) to test whether the two proportions are significantly different. Using Excel, we get a *CHITEST* probability of 0.0094, showing quite conclusive evidence of a

Advertisement	Number watching TV	Number aware of advertisement
Wednesday	216	121
Saturday	228	155

Figure 7.9   Advertising awareness data.

Advertisement	Number aware of advertisement	Number not aware of advertisement
Wednesday	121	95
Saturday	155	73

Figure 7.10   A two-way table of advertising awareness.

different level of awareness on the two days. For some reason it appears that the Saturday advertisement has had more impact.

Hence, it is probably inappropriate to combine the two days' results into a single figure for the overall impact of the advertisement. To be able to quote an overall awareness figure is appealing, but we must be careful when combining two (or more) sets of results that may be qualitatively different. For example, the viewing population might be different on the weekend than on weekdays. For instance, an older group could be watching TV on Saturday evenings, and if the advertisement has more resonance with an older audience, it could be expected to create greater awareness for that reason. Also, there could be a "carryover" effect from Wednesday, because the same viewers' panel is being questioned. The panel survey following the first screening may lead to an increased awareness and recall after the second screening a few days later.

7.13   Chapter Summary

In this chapter we have been concerned with estimating the population mean and population proportion based on a random sample of observations from the population. Specifically:

- The occurrence of "sampling error" that arises when using a random sample to estimate some population parameter.
- That bias may arise in nonrandom samples.
- Use of the margin of error to provide information about the accuracy of an estimate.
- The sampling distribution of the mean is approximately normally distributed, with the same mean as the population and a standard deviation that depends on the population standard deviation and the sample size.

- Percentage points of the normal distribution are used in the calculation of the margin of error for the mean if the population standard deviation is known.
- Percentage points of the *t*-distribution are used in the calculation of the margin of error for the mean when only the sample standard deviation is available.
- The normal distribution is used to determine the margin of error of a proportion.
- Testing whether the results from a sample are consistent with some hypothesised value of the population parameter is not very informative for decision making.

7.14 Exercises

1. The manager of a savings bank wants to estimate the average amount held in passbook savings accounts by depositors at her branch. A random sample of 25 depositors is taken. The sample mean is $4750 and the sample standard deviation $1200. Assuming that the amount in passbook savings accounts is normally distributed, calculate a 95% margin of error for the mean amount in passbook savings accounts.

2. The volume dispensed into soft drink bottles is known to be approximately normally distributed with a standard deviation of 0.05 litres. A random sample of 50 bottles gives a sample mean volume of 1.99 litres. Bottles that contain less than the specified volume (2 litres) can lead to prosecution of the manufacturer.
    a. What is the 90% margin of error for the mean volume? Comment on your results.
    b. The manager wants to estimate the mean volume to within 0.01 litres with 95% confidence. How many observations should be included in the sample?

3. A university wants to determine the average income its students earn during the summer.
    a. Explain briefly how you might choose a stratified sample of students.
    b. What factors would influence your choice of sample size?
        The university decided finally to take a random sample of 25 second-year management students. The mean and standard deviation of the income earned by these students was $3270 and $63.10, respectively.
    c. Calculate the margin of error for the mean summer income, with 99% confidence.
    d. What does this margin of error mean?
    e. Does this margin of error apply to all management students, or to all university students? Explain.

4. Ron Jones, the general manager of the National Paper Company (NPC), wants to determine the mean diameter of pine trees on land being considered for purchase. A random sample of 41 trees gives a mean of 41.42 cm with a standard deviation of 6.35 cm.
    a. Why do you think the mean diameter would be of interest to NPC?
    b. Construct a 95% margin of error for the mean diameter.
    c. Ron thought maybe he should have constructed a 99% margin of error. Explain briefly the effects of increasing the confidence level.

    d. Ron wishes to estimate the mean diameter to within 1 cm with 95% confidence. How many trees should be included in his sample?

5. A retail company hired an auditor, Francis Small, to verify the accuracy of its new invoice system. Francis randomly selected 70 invoices produced since the system was installed. She then compared each invoice against the relevant internal records to determine by how much the invoice was in error. She found that on average the error was $6.35, with a standard deviation of $35.30.

    a. Identify the population Francis was studying.

    b. She first calculates a 95% margin of error for the mean error per invoice. What value does she obtain?

    c. She concludes from this calculation that the margin of error does not make sense, because it was larger than the average error. Is her conclusion correct? Explain.

    d. Comment on the accuracy of the new invoicing system.

6. A bank manager is interested in the proportion of customers who have multiple accounts at the bank. A random sample of 500 customers is taken showing that 195 of these customers have multiple accounts.

    a. Calculate a 95% margin of error for the proportion of customers having multiple accounts.

    b. Describe in plain language what this means.

7. It is common practice, prior to most local and national elections, for the media to predict the winning party. Let us suppose that just two parties, Democrats and Republicans, are in an election. In a random sample of 353 voters, 205 (58%) said they intended to vote for the Democrats.

    a. What can you deduce from this information about the likely outcome of the election?

    b. How confident would you be that the Democrats will win?

8. The Healthy Breakfast Company carried out an advertising campaign on a new rice-based breakfast cereal, Extra Fruity. The manager would like to determine the proportion of students who had Extra Fruity for breakfast at least once during the week following the campaign. From a systematic sample of 200 students, he found that 18 of them had eaten this cereal for breakfast at least once in the week concerned.

    a. Construct a 99% margin of error for the proportion of students eating the cereal at least once in the week in question.

    b. The agency planning the campaign was targeted to achieve at least 15% penetration in the young consumer market. What do the sample results tell you regarding this objective?

9. A TV3-CM Research opinion poll puts support for NZ First above the margin of error for the first time in months, with 4.4% of 1000 voters polled last week saying they would vote for it. Discuss, showing any calculations you make.

10. A market researcher states she has 95% confidence that the true average monthly sales of a product are between $170,000 and $200,000. What does she actually mean?

11. As part of an audit investigation of a company, the auditors take a random sample of 150 invoices and find that 6 of these contain a financial error.

    a. What is the 90% margin of error for the proportion of all invoices containing errors? Explain what your answer means.

b. If the standard audit requirement is for a (90%) margin of error of no more than 2%, how large a sample should have been taken?

c. If the proportion of invoices containing errors could be as high as 10%, would this affect the size of sample required? If so, why?

d. Some people believe that a good sample size is about 10% of all invoices. Do you agree, and why?

12. During an audit, a random sample of 120 invoices is taken from the large number of invoices generated during the previous financial year. The mean value of these invoices is £357.45, with a standard deviation of £87.86. In addition, 6 of the invoices were for an incorrect amount or contained some other financial error.

a. Determine a 99% margin of error for the overall average invoice value. Explain carefully the meaning of this margin of error.

b. Determine a 90% margin of error for the overall proportion of incorrect invoices.

13. Jason is the manager of a stationery shop. Over the past 5 years he has noticed that the standard deviation of the price of the greeting cards he sells is about 60 cents, even if the average price changes. He takes a random sample of 20 cards from the shelf and calculates the sample mean price to be $3.85.

a. Calculate a 90% confidence interval for the mean price of his stock.

b. Give a reason why this might not be a realistic estimator of the average value of cards sold.

c. How would you advise Jason to do his sampling in order to get a more effective estimate of the average value of cards sold? Describe both the sampling frame and the sampling method.

14. Alan Jones, the manager of Fine-Fare supermarket in Leicester, is looking into the viability of opening on Sundays. His daily sales on weekdays averages £120,000, with a standard deviation of £8,000. Taking into account wages and other store overhead, Brian decides that if the supermarket will gross at least £50,000 on a Sunday, then it will be worth opening. He decides to open on a trial basis for the next six Sundays, when his daily sales (in thousands of pounds) are:

$$64, 58, 49, 55, 71, 51$$

a. Calculate the mean and standard deviation of daily sales over these six Sundays.

b. Determine a 95% margin of error for the expected average Sunday sales. What would be your advice to Alan regarding whether he should open on Sundays?

c. If you could assume that the standard deviation of weekday sales (i.e., £8000) is also appropriate for Sunday sales, would this make a difference to the margin of error and to the advice you give to Alan?

15. Frobisher Whiteware Appliances Inc. is concerned about how many of its most popular model of washing machine, the Tender Spin, need repair within the 24-month warranty period. Walter Arnold, the quality control manager, believes this could be as high as 20%. To investigate this problem, Walter decides to take a random sample from all machines that were sold between two and three years ago and find out how many have been repaired under warranty.

    a. Advise Walter on how large a sample size he should take in order to estimate the proportion of all machines sold that are repaired under warranty, with a 95% margin of error of no more than 5%.

    b. He selects the size of sample you recommend and finds that 38 of these machines have been repaired under warranty within the first two years. Determine a 95% margin of error for the proportion of all machines sold that are repaired under warranty.

    c. Is it likely that the proportion could be as high as Walter originally thought?

16. Following a proposal to rename the city of Hamilton, one of the weekly papers decided to carry out an opinion poll. They selected every 10th page from the local telephone directory and called the 23rd residential number listed on each page. Information was collected from whoever answered the phone, provided the individual was over 15 years of age. A total of 268 numbers were called, of which 217 calls were answered. Of these, 26 people under 15 answered the call, and a further 31 refused to give any information.

    a. What type of sample is being used?

    b. In estimating the proportion of people likely to vote for each of the possible alternative city names, what is the maximum margin of error that could arise?

    c. Why does this margin of error need to be associated with a particular "level of confidence"?

    d. A total of 23 people chose River City as the new name. Explain why the margin of error you calculated in (b) does not apply here, and calculate a more accurate margin of error for the proportion of people choosing River City.

    e. Explain carefully what this margin of error means.

17. Michael Young, the internal auditor of the St. Lucia Trading Company, carried out an investigation into the accuracy of the company's new billing system. He took a random sample of 120 invoices produced under the new system. He then compared each invoice against the relevant internal records to determine how much the invoice was in error. He found that the average error was $1, with a standard deviation of $83.

    a. Identify the population Michael studied.

    b. Construct a 95% confidence interval for the mean error per invoice.

    c. Interpret this confidence interval.

    d. What conclusions should Michael draw from his analysis, and what action should he now take?

Chapter 8

Regression Analysis

8.1 Introduction

In this chapter and the next, we consider a variety of situations where valuable information can be obtained by investigating relationships between variables. This is the topic of *regression analysis*. Regression is widely used in business, but it is also widely misused. It is very easy to become so involved with the complexities of the various regression formulae and with endless calculations that the purpose and scope of regression are overlooked. As usual we shall leave the calculations to Excel and focus instead on what regression means, the basic ideas involved in carrying out a regression analysis, checking whether the underlying assumptions have been satisfied, and how to use the results.

We start by considering five data sets and asking you some questions about them. This should give you some insight into regression and what should be done next.

Ice Cream Sales

Wiremu Harvey, the marketing manager of the New Top Ice Cream Company, is reviewing the company's sales data. The sales measured over 30 four-week periods, together with the mean daily temperature in each period, are given in Figure 8.1.

> *Q1. How might Wiremu use these data?*
> *Q2. What questions should he be asking?*

The first thing to do when studying the relationship between two quantitative variables is to plot the data. *Failure to plot the data often leads to incorrect*

Period	Sales (kilolitres)	Mean temp (°C)	Period	Sales (kilolitres)	Mean temp (°C)	Period	Sales (kilolitres)	Mean temp (°C)
1	438	5	11	325	-2	21	362	7
2	425	13	12	339	-3	22	349	4
3	446	17	13	374	0	23	323	0
4	483	20	14	361	4	24	370	-3
5	461	21	15	433	13	25	351	-2
6	391	18	16	433	17	26	408	1
7	371	16	17	534	22	27	427	5
8	327	8	18	503	22	28	473	11
9	306	0	19	438	19	29	496	18
10	291	-4	20	389	16	30	523	22

Figure 8.1   Ice cream sales and the mean daily temperature.

analysis and to inappropriate predictions. A scatterplot is used to plot such data, with one variable plotted on the vertical axis (the *y*-axis) and the other variable on the horizontal axis (the *x*-axis). Scatterplots were discussed in Section 3.13.

The scatterplot of the ice cream sales data in Figure 8.1 is given in Figure 8.2.

Q3. *How would you use this plot to predict sales from an estimated mean daily temperature? What do you estimate sales would be when the mean daily temperature is 10°C? when it is 20°C?*

Q4. *How would you describe the relationship between sales and temperature?*

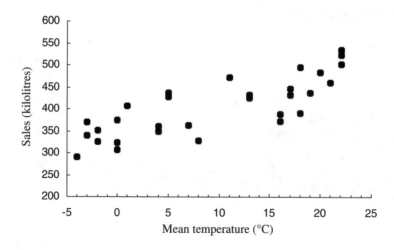

Figure 8.2   New Top ice cream sales.

That companies are using data, such as that in Figure 8.1, on daily temperature to help predict sales can be seen from the following newspaper extract:

> British companies are tuning in to weather forecasts to help predict consumer trends — a move that is saving them millions of pounds a year.... We are analysing historical sales statistics...and identifying the weather factors which influenced those sales.... Mathematical equations are then devised to pinpoint the precise relationship between weather and sales to enable retailers to predict demand much more accurately. This method [is used] to predict fluctuations in sales of a single brand such as Unilever's Persil detergent (people do more washing when the weather is fine) or Nestle's Kit-Kat chocolate bars, which can go soft in hot weather.... The weather also has a big impact on the purchase of toilet rolls and can account for monthly variation in sales of up to 20,000 packs. (Source: *New Zealand Herald* [Reuters], August 5, 1997.)

Raw Material Moisture Content

During a certain production process, it is necessary to decrease the moisture content of some intermediate product. A cause and effect diagram has revealed that one potential problem is the moisture content in the raw material. Figure 8.3 gives 50 pairs of values recorded on the percentage moisture content of the raw material (RM) and the moisture content of the intermediate product (PROD) made from it. (Adapted from Ishikawa's *Guide to Quality Control,* Asian Productivity Organisation, 1982). Figure 8.4 shows a scatterplot of the data.

No.	RM	PROD	No.	RM	PROD	No.	RM	PROD	No.	RM	PROD	No.	RM	PROD
1	1.10	1.40	11	1.50	1.40	21	1.30	1.50	31	1.85	2.10	41	1.05	1.85
2	1.25	1.70	12	1.40	1.50	22	1.45	1.55	32	1.40	2.00	42	1.35	2.10
3	1.05	1.30	13	1.35	1.70	23	1.20	1.55	33	1.60	2.30	43	1.60	2.10
4	1.75	1.75	14	1.20	1.40	24	1.90	1.90	34	1.10	1.60	44	1.45	2.20
5	1.30	1.30	15	1.00	1.35	25	1.65	1.70	35	1.60	1.75	45	1.20	1.80
6	1.15	1.20	16	1.40	1.30	26	1.05	1.85	36	1.85	2.40	46	1.35	1.80
7	1.70	1.40	17	1.60	1.60	27	1.60	2.05	37	1.70	2.30	47	1.05	1.70
8	1.60	1.95	18	1.50	1.85	28	1.55	2.30	38	1.55	1.90	48	1.30	2.30
9	1.50	1.50	19	1.80	1.70	29	1.40	2.00	39	1.45	2.15	49	1.45	1.80
10	1.80	1.90	20	1.65	1.55	30	1.30	1.90	40	1.15	2.00	50	1.30	1.70

Figure 8.3   Raw material and product moisture content.

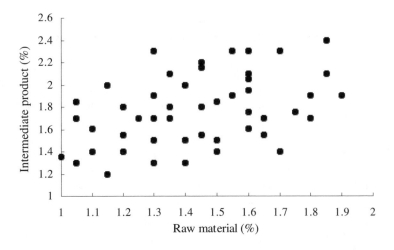

Figure 8.4   Product and raw material moisture content.

Q5. What question might the production manager be interested in asking?
Q6. What can you say about the relationship between the percent moisture
contents of the intermediate product and the raw material?

Inflation and Interest Rates

The data in Figure 8.5 give the average annual rates of interest and inflation between 1979 and 1992. A scatterplot of the data is shown in Figure 8.6.

Year	Inflation rate (%)	Interest rate (%)	Year	Inflation rate (%)	Interest rate (%)
1979	6.9	17.2	1986	3.9	9.0
1980	6.5	16.4	1987	2.5	8.3
1981	6.3	14.5	1988	3.1	7.7
1982	6.1	12.5	1989	2.4	7.5
1983	5.5	11.3	1990	1.9	6.9
1984	5.0	10.2	1991	1.3	6.2
1985	4.3	8.7	1992	0.8	6.0

Figure 8.5   Inflation and interest rates.

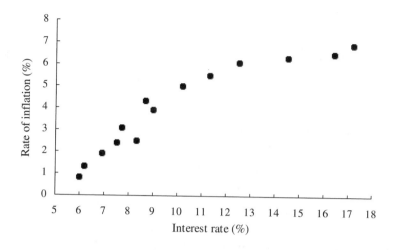

Figure 8.6   Interest and inflation rates, 1979–1992.

Q7.   Why do you think a relationship between annual interest and inflation rates might be useful to know?

Q8.   How would you describe the relationship between interest rates and inflation rates over the period 1979–1992?

📈 A Salary Survey

In a salary survey of its members, the National Society of Accountants collects data on salary (in thousands of dollars), age, and qualifications. Each member's qualifications are converted into a score (QS), where the higher the score, the greater the qualifications possessed. Figure 8.7 shows a sample of data from 30 members. Two scatterplots of the salary survey data are given in Figure 8.8. The first plots salary against age, and the second salary against qualification score.

Q9.   How do you think the Society might use these data? What questions might be of interest?

Q10. What would you conclude from these plots? What other plots might you wish to see?

Salary	Age	QS	Salary	Age	QS	Salary	Age	QS
372	68	14	164	49	19	219	42	25
206	58	17	113	45	19	186	33	24
154	52	17	82	34	20	155	31	25
175	47	14	32	32	22	114	24	27
136	42	17	228	57	20	341	62	24
112	42	17	196	45	23	340	54	25
55	32	18	128	38	21	283	48	25
45	59	19	97	30	24	267	39	26
221	60	20	64	25	28	215	32	27
166	53	21	249	54	24	148	27	27

Figure 8.7   Salary data from the National Society of Accountants.

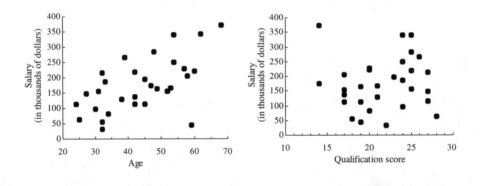

Figure 8.8   Salary against age and qualification score.

📖 *Noise Levels at London Gatwick Airport*

A study was conducted at London Gatwick Airport to investigate the existing procedures for the prediction of aircraft noise. The aim was to predict the perceived noise level (PNL) given the slant distance in metres (SD), which is the distance from the point at which the aircraft starts its takeoff to its position when it passes over the noise recorder located beyond the end of the runway. The data collected are shown in Figure 8.9, with a scatterplot given in Figure 8.10.

> **Q11.** *Look carefully at the data. What, if anything, do you notice?*
> **Q12.** *What PNL value would you predict for a slant distance of 600 metres?*
> **Q13.** *What conclusions do you draw from the scatterplot?*

No.	SD	PNL	No.	SD	PNL	No.	SD	PNL	No.	SD	PNL	No.	SD	PNL
1	993	107	13	982	99	25	246	114	37	195	127	49	1000	98
2	1019	98	14	248	117	26	192	117	38	213	117	50	342	116
3	977	102	15	204	120	27	1037	97	39	37	119	51	476	115
4	182	120	16	149	120	28	169	124	40	400	115	52	1029	100
5	295	114	17	207	116	29	36	131	41	298	120	53	1083	99
6	96	123	18	211	116	30	86	128	42	455	112	54	293	121
7	93	121	19	1037	100	31	250	118	43	328	117	55	214	123
8	994	100	20	178	115	32	375	122	44	1014	104	56	1008	98
9	136	121	21	207	115	33	1046	102	45	1072	107	57	996	99
10	204	119	22	248	111	34	1007	97	46	1024	106			
11	1015	97	23	1008	91	35	991	100	47	328	117			
12	996	101	24	205	118	36	250	121	48	234	120			

Figure 8.9   Noise levels at Gatwick Airport.

The two sets of points represent aircraft that take off steadily (a low slant distance) or sharply (a high slant distance). Plotting the two sets individually gives the scatterplots in Figure 8.11.

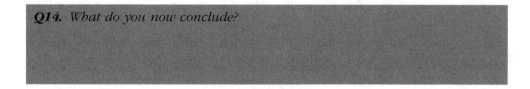

Q14. What do you now conclude?

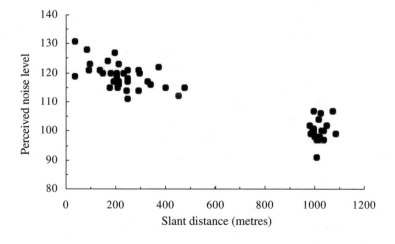

Figure 8.10   Gatwick Airport data.

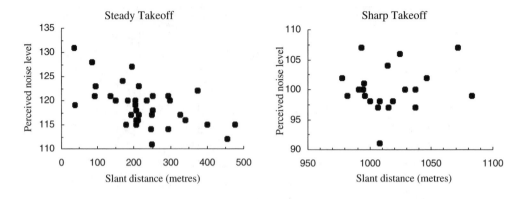

Figure 8.11   Scatterplots for steady takeoffs and sharp takeoffs.

This example illustrates very clearly the potential dangers of trying to predict without first plotting the data. In fact, such a comment is true for any form of analysis.

8.2   Relationships between Variables

We have looked at five examples where we examined the relationship between two quantitative variables. In the ice cream example the marketing manager may be interested in trying to predict sales from mean daily temperature. This could be of considerable value if reliable weather forecasts are available. Of course, temperature is unlikely to be the only variable that could be used to predict ice cream sales. We shall consider later how two or more variables might be used jointly to predict sales.

In the moisture content example, if there is a strong relationship between the moisture contents of the raw material and the intermediate product, then it may be possible to control the moisture content of the intermediate product by ensuring that the moisture content of the raw material is kept at certain values. For example, in New Zealand sawmills, radiata pine is dried in kilns. The moisture content of the dried wood is a critical factor in further processing and is known to be strongly related to the moisture content of the timber before it goes into the kiln. This type of relationship can be depicted as in Figure 8.12 and can often be exploited to control the outputs.

Another situation where knowledge of the relationship between two variables can be very valuable is when the variable of interest either takes too long to measure or is expensive to measure. If a substitute variable can be found

Figure 8.12   Relationship between inputs and outputs.

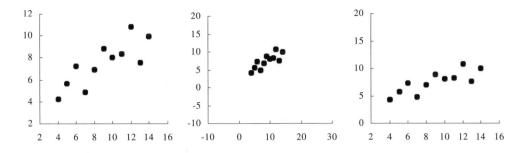

Figure 8.13   Effect of changing scale on a scatterplot.

that is strongly related to the variable of interest, then this substitute variable (sometimes called a *proxy variable*) can be used instead. For example, the amount of shrinkage in an injection-moulded plastic product is usually critical. But it takes a few hours for the product to cool sufficiently so that a final measure of shrinkage can be obtained, by which time it will too late to make any adjustments to the injection-moulding machine. However, if the relationship between final shrinkage and shrinkage at the time of production is known, then an immediate estimate of shrinkage can be obtained.

As we have seen in the previous section, drawing a scatterplot should be the first step when looking for a relationship between two variables. A scatterplot will indicate whether the expected relationship actually exists and throw some light on the form of any relationship. It may also reveal important features in the data that should be taken into account when analysing apparent relationships. For example, there could be outliers in the data, such as those in Figure 8.8, or the data could cluster into clear groups, as in Figure 8.10. However, care has to be taken when using scatterplots to examine relationships between variables. More space around the data can make a relationship look stronger. Changing the spread of one variable can change the perceived strength of the relationship. All three scatterplots in Figure 8.13 have been obtained from the same set of data, but you might think that one relationship looks stronger than the others.

8.3   Regression Analysis

Regression analysis is widely used in management for determining the form of the relationship between two or more related variables. However, it is also one of the most misused statistical techniques. Our main aim in this chapter is to acquaint you with the role, scope, and interpretation of regression. Knowing when and when not to use regression is more important than understanding the technical details of calculating regression equations.

First consider a simple linear relationship involving two quantitative variables; the *response variable* and the *explanatory variable*. One of the primary purposes of regression analysis is to predict the value of the response variable for a given value of the explanatory variable. In a scatterplot it is usual to put the response variable on the vertical axis, or *y*-axis, and the explanatory variable on the horizontal axis, or *x*-axis. For this reason the response variable is some-times referred to as the *y-variable* and the explanatory variable as the *x-variable*. In many books the response variable is called the *dependent* variable and the explanatory variable the *independent* variable.

One common mistake in regression analysis is to confuse the two types of variables. Sometimes either could be the response variable; in other cases only one of them can sensibly be used. The issue to keep in mind is "which variable are we interested in predicting, given a value of the other variable." For example, if the two variables are the heights and weights of a group of schoolchildren, then either variable could be the response variable, and which we choose depends on the purpose of the study. On the other hand, in the ice cream sales example, the response variable is clearly the sales data, because predicting daily temperature given sales of ice cream makes no sense.

Many books do not make it clear that when they talk about regression analysis they really mean *simple linear* regression. That is, they are concerned with fitting *straight lines* to data. This may be totally inappropriate, for a number of reasons, but principally when the evidence of the scatterplot suggests that the relationship is nonlinear, as in Figure 8.6. Failure first to plot and examine the data probably leads to the most frequent misuse of regression analysis.

Regression Analysis of Ice Cream Data

Consider again the ice cream sales data in Figure 8.1. The starting point of any analysis is the scatterplot, which is reproduced in Figure 8.14.

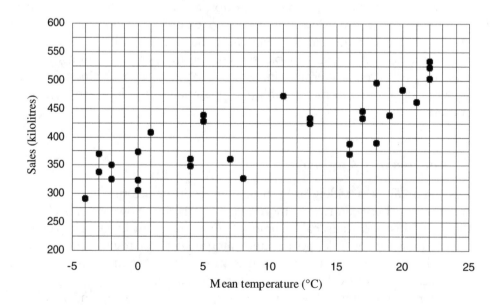

Figure 8.14   New Top ice cream sales.

To illustrate regression analysis we will first develop a *subjective* relationship between ice cream sales and mean daily temperature:

1. Draw a straight line on the plot, which you think best "fits" these data points.
2. Read off the value of ice cream sales where your line crosses the vertical axis *at a temperature of 0°C.* We shall denote this *intercept* by *a*. Hence, *a* = _____.
3. Calculate the slope of the line. That is, calculate from your line the change in sales resulting from a 1°C change in temperature. To do this, read off the sales at 20°C, work out the change in sales from 0°C to 20°C, and divide this by 20. The *slope* will be denoted by *b*. Hence, *b* = _____.
4. The equation of your straight line is

$$y = a + bx = \text{\underline{\hspace{4cm}}}$$

where *y* is the response variable (sales) and *x* is the explanatory variable (mean temperature).

8.4 Least Squares Method

In practice, fitting a model by eye is too imprecise and can lead to substantial differences from one person to another. We need a precise criterion for determining objectively the equation of the *line of best fit* for any set of data. This is the *least squares* criterion.

Again we shall illustrate the procedure with the ice cream sales data. The scatterplot is reproduced in Figure 8.15. We have also drawn in a straight line that best fits the data according to the criterion we now describe. We need to estimate the intercept and slope of the line so that the model fits these data

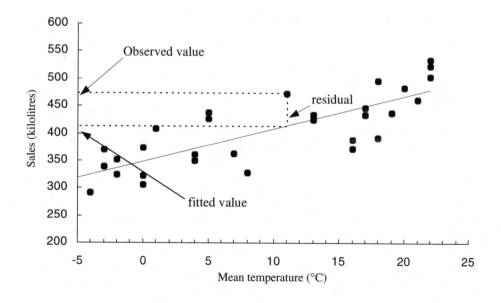

Figure 8.15 New Top ice cream sales.

well. As you will see in the next section, the least squares line of best fit to the data has the equation

$$y = 348.9 + 5.9x$$

For the value $x = 11°C$, the ice cream sales predicted by the regression line is $348.9 + 5.9 × 11 = 413.8$; this is called the *fitted value* of y. The discrepancy between what we observed and the fitted value is called the *residual* of y (or the *error* in the fitted value). The fitted value and residual corresponding to the observed value at 11°C are shown in Figure 8.15.

Thus residuals are defined as

$$\text{Residual } (e) = \text{Observed value of } y - \text{Fitted value of } y$$
$$= y - (a + bx)$$

Any observed value of y is, therefore, represented by a *model* that may be written symbolically as

$$y = a + bx + e$$

which expresses the observed value as a "discrepancy" (e) from a linear trend with intercept a and slope b. To fit a regression line by least squares, we choose a and b to minimise the sum of squares of these residuals.

Apart from being the basis for least squares analysis, residuals provide important information for examining both the adequacy of the fitted model and the assumptions underlying our analysis. The study of residuals is a vital part of any regression analysis, and we shall look more closely at them and the assumptions we are making in Section 8.6.

✍ *Least Squares Demonstration*

To illustrate the process of fitting a regression line by least squares we can use the Excel spreadsheet *Regression.xls* provided on the CD-ROM. The first sheet gives instructions on how to use this spreadsheet and the second sheet gives the demonstration. The data shown on the scatterplot in the second sheet are from the ice cream sales example given in Figure 8.1. Initially the scatterplot shows a horizontal line at the mean level of sales (405). This is depicted in Figure 8.16. At the bottom left of the screen are the intercept (405) and slope (0) of the line. Above each are sliders that allow the intercept and slope to be changed in increments of 1 and 0.1, respectively. The residual sum of squares is calculated as the sum of squares of the vertical distances of each point from the line, shown by the vertical lines connecting the data points to the line. Initially this has a value of 128,606.0.

You can use the sliders to change the slope and intercept to find the line that gives the lowest possible residual sum of squares. Give it a go! To the right of the graph is a tube filled with yellow "liquid," which represents the residual sum of squares. As the residual sum of squares decreases, the tube starts

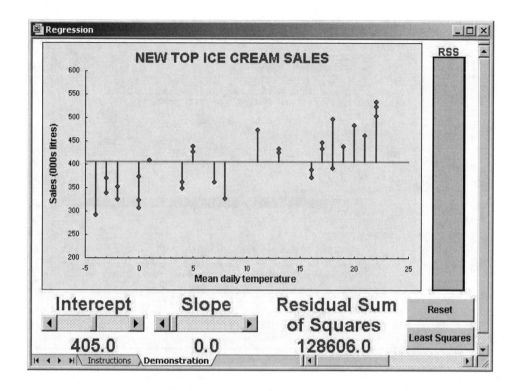

Figure 8.16   Regression demonstration spreadsheet.

emptying. When you have reduced the residual sum of squares to its minimum value, the liquid will have completely disappeared.

When you think you have found the lowest residual sum of squares, click on the *Least Squares* button to get the line that gives the least squares fit to the data. You will see that this line gives a residual sum of squares of 45,550.7.

If you want to have another go, click the *Reset* button to return the graph to its initial position.

✎ *Calculating the Regression Line Using Excel*

In most statistics books you will find formulae for calculating the least squares slope and intercept, but this calculation is tedious and time consuming to do by hand. Instead we can calculate the slope and intercept with Excel, using one of the following methods.

First we input the data into a spreadsheet. In the Excel spreadsheet in Figure 8.17, we have put the ice cream sales data (y) into cells A1:A30 and the mean daily temperature (x) into cells B1:B30. Only the first 20 values of the two variables are shown here. The *Insert Function* command can be used to calculate both the intercept (a) and the slope (b) by choosing *Statistical* in the function category and selecting either *INTERCEPT*, as shown in Figure 8.17, or *SLOPE* in

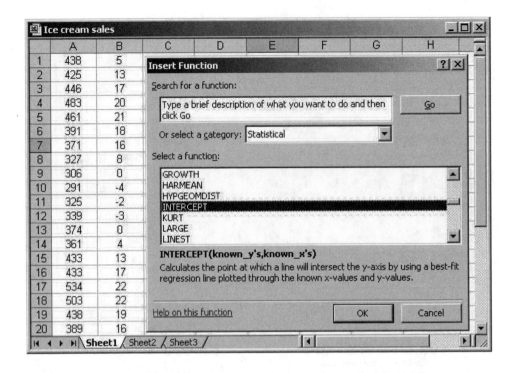

Figure 8.17   Excel spreadsheet windows for calculating the intercept.

the function name. This opens a dialogue window where the ranges containing both the *x*- and *y*-values are specified and from which the intercept or slope is calculated. Figure 8.18 shows the *INTERCEPT* window; the equivalent *SLOPE* window is effectively identical.

A second method is to use the regression facilities in the *Data Analysis Tools Add-In*. Choosing *Regression* from the *Data Analysis* menu gives the dialogue box in Figure 8.19. The data ranges for *x* and *y* are entered into the appropriate *Input* boxes. If our spreadsheet contains names for the *x* and *y* variables at the head of each range, then these cells could be included in the data ranges and the *Labels* box checked, in which case the output will refer to the variables by name. None of the other dialogue options are required at this stage, although later we shall use some of them. Part of the output from this analysis is given in Figure 8.20.

The fitted regression equation is therefore

$$y = 348.9 + 5.9x$$

or

$$\text{Sales} = 348.9 + 5.9 \times \text{Mean daily temperature}$$

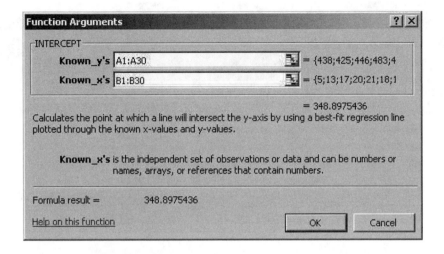

Figure 8.18   *INTERCEPT* dialogue window.

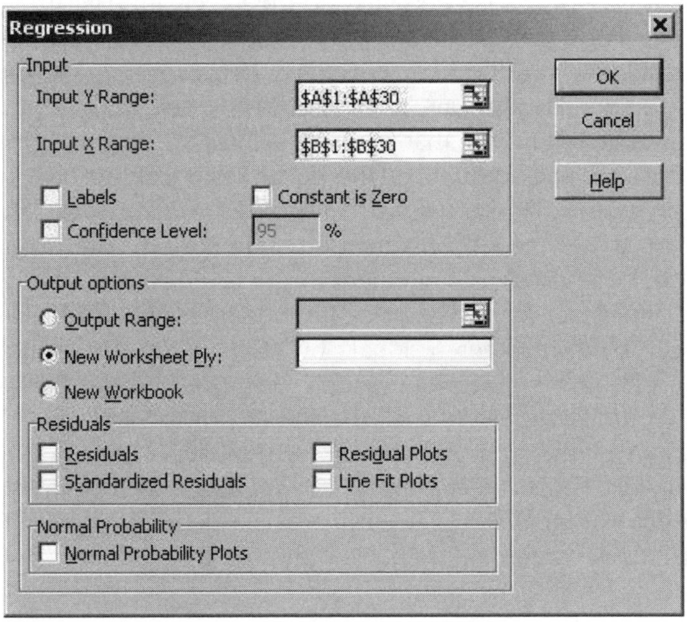

Figure 8.19   Regression data analysis dialogue window.

Regression Statistics			Coefficients
Multiple R	0.8036	Intercept	348.898
R Square	0.6458	X Variable 1	5.906
Standard Error	40.3338		
Observations	30		

Figure 8.20   Selected regression output.

> **Q15.** *For the ice cream data, the intercept is calculated as 348.9, as shown in Figure 8.18 or from the output in Figure 8.20. What does this intercept of 348.9 mean?*
>
> **Q16.** *In a similar way, using the SLOPE function or from the output in Figure 8.20, the value of the slope is given as 5.9. What does this value mean?*
>
> **Q17.** *Given this regression line, what would be your predicted sales for a mean temperature of 10°C?*

8.5   Strength of the Relationship

The line $y = 348.9 + 5.9x$ has been fitted to the ice cream data using regression. Is it a good fit? For a perfect fit the line would go through all the data points and all the residuals would be equal to zero. However, in most practical situations the line looks like the one in Figure 8.15, where most of the points do not lie on the regression line; that is, the residuals are mainly nonzero.

We can measure the strength of the linear relationship between ice cream sales and temperature by the residual sum of the squares. The larger this sum of squares, the worse the fit. This measure suffers from one major drawback: It depends on the units of measurement of the response variable. For instance, if we were to measure ice cream sales in litres rather than thousands of litres, we would get a much larger residual sum of squares. Thus, the size of the residual sum of squares depends not only on how well the line fits the data, but also on the units we use to measure the response variable.

To see how well the least squares line fits the data, we need a measure of fit (a *coefficient*) that is not affected by how we choose to measure the response variable. Ideally, we need a coefficient that has a defined range, such as 0 to 1, where the larger the value, the better the fit. The measure of fit used, R^2, does just this.

✏ *Calculating and Interpreting* R²

The least squares regression demonstration showed that the residual sum of squares is 128,606.0 when the fitted line had an intercept equal to the mean of the response variable and a slope of zero. It represents the total variation in the sales data. We also saw that the minimum residual sum of squares is 45,550.7 for the least squares regression line, which represents the amount of variation in sales still unexplained by the explanatory variable temperature. Therefore, in comparison with a horizontal line, the least squares line reduces the residual sum of squares by 128,606.0 − 45,550.7 = 83,055.3. The proportion that has been explained by temperature is the coefficient R^2:

$$R^2 = \frac{83055.3}{128606.0} = 0.646$$

In other words, the linear regression model explains 64.6% of the variation in ice cream sales with temperature as the explanatory variable.

If there were no linear relationship, the initial residual sum of squares could not be improved by fitting a different line, and so $R^2 = 0$. If all the points fall exactly on a line with nonzero slope, the residuals will all equal zero, and hence we have a residual sum of squares of 0. Thus the entire initial residual sum of squares has been removed, giving $R^2 = 1$.

We can calculate R^2 in Excel using the *Statistical* function *RSQ*. For the ice cream data, with sales and mean temperature in cells A1:A30 and B1:B30, respectively, we get an R^2 value of 0.6458, as before. The dialogue window for *RSQ* is shown in Figure 8.21. It is also given as part of the *Regression Data Analysis* facility, shown as *R Square* in Figure 8.20.

✍ *The Correlation Coefficient*

Closely related to R^2 is the *correlation coefficient r*, which also measures the linear relationship between two variables. *Multiple R* given in Figure 8.20 is the square root of *R Square*. If a positive linear relationship exists between two variables (positive slope), then $r = +R$; if a negative linear relationship exists (negative slope), $r = -R$. The correlation coefficient is calculated in Excel using the statistical function *CORREL*. The dialogue window for *CORREL* is effectively identical to that for *RSQ* shown in Figure 8.21.

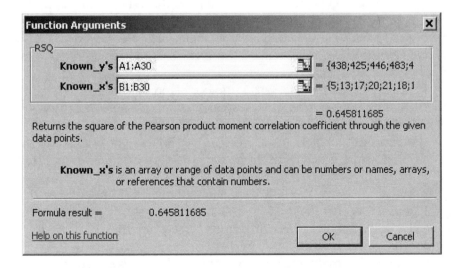

Figure 8.21 Dialogue window for the *RSQ* function.

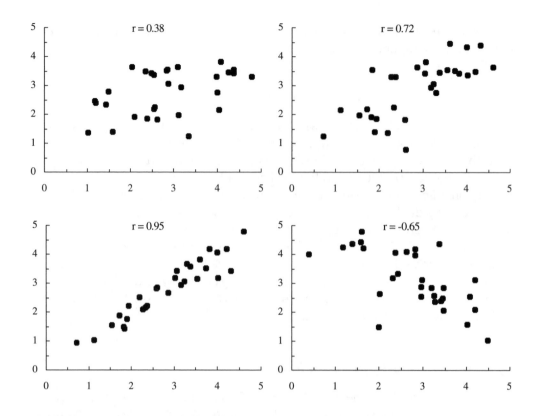

Figure 8.22   Scatterplots and the correlation coefficient.

As with R^2, the correlation coefficient only measures *linear* association between two variables. It is important, therefore, always to construct a scatterplot of the data. If you observe a linear relationship, then r is another good summary of the strength of that relationship. If the relationship is nonlinear, then it should be apparent from the scatterplot.

The scatterplots in Figure 8.22 give you some idea of the strength (strong or weak) and sign (positive or negative) of the correlation coefficient for different scatterplots.

For the ice cream data in Figure 8.1, the correlation coefficient between sales and mean temperature is +0.804, indicating a strong positive linear association between these two variables.

For the moisture content data in Figure 8.3, the correlation coefficient between the moisture contents of the intermediate product and raw material is +0.397.

The correlation coefficient between inflation rates and interest rates for the data in Figure 8.5 is +0.933. Although this indicates a strong linear relationship between the two variables, we can see from the scatterplot that a curvilinear relationship is even stronger, as we will show in Chapter 9.

For the Gatwick Airport example in Figure 8.11, the correlation coefficient between slant distance and noise level is –0.564 for the steady takeoff and

+0.145 for the sharp takeoff. The latter case presents little evidence of any linear relationship between the two variables, which is confirmed by the scatterplot.

8.6 Analysing the Residuals

A scatterplot of the data should be examined before any analysis is carried out. Often this will tell you whether a linear regression is likely to give a good fit to the data. In addition, an analysis of the residuals can give information not only about the adequacy of the model but also about an assumption we need to make about the distribution of the response variable.

From the fitted regression model we usually want to predict the value of the response variable y given a value of the explanatory variable x. For example, you can predict what ice cream sales will be for a given mean daily temperature. For instance, at a temperature of 0°C you should get sales of about 350 (i.e., 350,000 litres). Of course few, if any, of the actual values of y at $x = 0$ will be exactly 350. Looking at the data in Figure 8.1 we see that we have three readings at 0°C: 306, 374, and 323. This is quite a bit of variation. If we had a lot more data, we might have other readings at 0°C. With masses of data we could build up a distribution of sales figures at 0°C. The fitted value from the regression line is the mean value of this distribution. The same applies to other temperature values.

So we can picture a distribution of sales values (y) at any level of temperature (x), whose mean value is given by the fitted regression model. What about the amount of variation in these distributions? When we carry out a regression analysis we make the following important assumption:

> *The standard deviation of the response variable is the same at any value of the explanatory variable. That is, the spread of the distribution of* y *is the same for all values of* x. *Only the mean value of* y *changes for different values of* x.

The standard deviation of these distributions can be estimated from the sample data, and it is part of the standard Excel regression output. It is calculated by dividing the residual sum of squares by the number of degrees of freedom and taking the square root. However, unlike earlier standard deviations, this *residual standard error* (s_e) has only $n - 2$ degrees of freedom, not $n - 1$. This is because the regression line from which the residuals are calculated requires the estimation of *two* parameters, the intercept and the slope. Hence, *two* degrees of freedom are lost. In the ice cream data the residual sum of squares is 45,550.7, so $s_e = \sqrt{45550.7 / 28} = 40.3$. This is the value shown as the *Standard Error* in Figure 8.20.

Careful examination of the residuals is important before we use the standard error s_e. If the model is inappropriate or the residuals are unusually large, then s_e will overestimate the true standard deviation, and any margins of error we calculate will be too large.

✎ *Calculating Residuals*

A residual is simply the difference between any observed value of y and the corresponding fitted value determined from the least squares regression line. For example, the first period in the ice cream data had observed sales of 438 when the mean temperature was 5°C, so

$$\text{Fitted value} = 348.9 + 5.9 \times 5 = 378.4$$

and

$$\text{Residual} = 438 - 378.4 = 59.6$$

Q18. What are the fitted values and residuals for the second and third periods?

In practice, we can use the *Regression Data Analysis* facility in Excel to give us the residuals, and a residual plot. For the ice cream data the Regression dialogue window is shown in Figure 8.19. To obtain the residuals in addition to the standard regression output, we simply check the *Residuals* box. This produces a list of residuals, part of which is reproduced in Figure 8.23. Note that the *fitted value of y* is referred to here as the *Predicted Y*.

RESIDUAL OUTPUT

Observation	Predicted Y	Residuals
1	378.43	59.57
2	425.67	-0.67
3	449.29	-3.29
4	467.01	15.99
5	472.91	-11.91
6	455.20	-64.20
7	443.39	-72.39
8	396.14	-69.14
9	348.90	-42.90
10	325.28	-34.28
:	:	:

Figure 8.23   New Top ice cream sales — sample residual output.

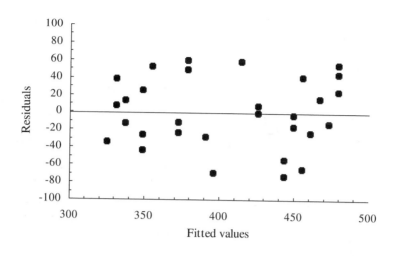

Figure 8.24   New Top ice cream sales — residual plot.

✍ *Examining the Residuals*

A simple way of examining the residuals is to draw a scatterplot of the residuals (on the *y*-axis) against the fitted values (on the *x*-axis). This should show a random scatter about the *x*-axis, with approximately equal numbers of positive and negative residuals. Such a plot for the ice cream sales data is given in Figure 8.24.

> *Q19. Identify the points corresponding to periods 1, 2, and 3 in Figure 8.24.*
> *Q20. What does this plot tell you about the linear model we have fitted?*

A clear pattern in a plot of the residuals against the fitted values suggests that either of the following is true:

■ The assumption of a constant standard deviation does not hold. For instance, a funnel effect, with greater variation in the residuals as the fitted value increases, suggests that larger values of *y* have greater variation. This is depicted in Figure 8.25. Fitting a different model to the data or using log (*y*) or the square root of *y* instead of *y* may overcome these problems.

■ The model is inappropriate. For instance, consider the inflation data in Figure 8.5. A scatterplot of the original data and a residual plot are shown

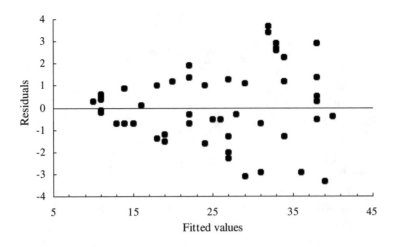

Figure 8.25   Residual plot.

in Figure 8.26. We can see from the scatterplot on the left that a straight line is not the most appropriate model for the data. The curved pattern in the residual plot on the right confirms this. One possible solution is to extend the regression model to include a quadratic term to account for this curvature. We shall see how to do this in Chapter 9.

In a two-variable regression, the residual plot merely confirms what should be evident from the scatterplot. However, in complex situations involving more than two variables, it is difficult to draw a comprehensive scatterplot, and the residual plot is the simplest way of checking the suitability of the model, as will be shown in Chapter 9.

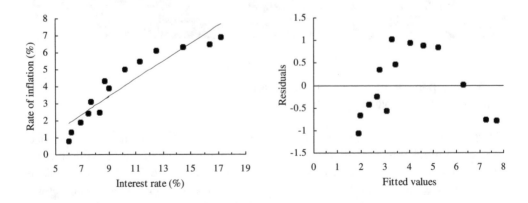

Figure 8.26   Scatterplot and residual plot for inflation rate data.

✎ *Other Residual Plots*

Plot of Residuals against the Explanatory Variable

If the model is linear in the explanatory variable, this plot should be randomly distributed about the x-axis. A curved pattern, as in Figure 8.26, suggests that the relationship is nonlinear in x. If you have several explanatory variables (see Chapter 9 on multiple regression), it is important to do this plot for all available explanatory variables, whether or not you have included them in the model.

Run Chart of Residuals in Time Order

It can be useful to plot the residuals against the time order in which the data were collected. This may alert you to any relationships between observations that are close in time. Cyclical behaviour suggests that the mean of the process is changing over time. Zigzag behaviour is often seen when compensatory behaviour is present. For example, a product is being used at a constant rate, but we ordered too much this month. So we order less next month. Splitting, or stratifying, the data according to some other variable may help you understand the variation better.

8.7 Unusual Observations

Observations that do not quite belong have always been a bother for data analysts. We call such unusual observations *outliers*. It is tempting just to get rid of them. But this is a shortsighted way to proceed, equivalent to ignoring anything that violates our assumptions. We need to know the reasons behind things in order to be able to understand and then improve them. Sometimes the outlier is an important discovery, such as a previously unknown feature of the relationship between the variables.

However, we should first check each unusual observation carefully. Is it simply a mistake, for example, typing 347 instead of 34.7? Is it an impossible result, such as a machine out of order for 26 hours on Thursday? We need to check our data as soon as possible so that we can find out whether mistakes have been made. We correct the data where possible and leave out false or incorrect numbers.

If we are satisfied an unusual observation is not a mistake, then we need to ask whether this observation affects the results unduly. One way to determine this is to reanalyse the data with the unusual observation omitted. The model will probably fit better, since we took the unusual part away. But how is the model different? Are the parameters (intercept and slope) different? Is there significantly less variation in the residuals?

- ■ If it makes little difference whether the unusual observation is in or out, then leave it in.

■ If it does make a difference then:
  • You need to ask whether the model with the unusual observation omitted now appears to fit the data well, in which case you can report that the model fits for all data except this unusual one, and think why it is unusual.
  • If a linear model still does not fit, you may need to try describing the relationship in another way.

8.8 Prediction from Regression

As we saw in Section 8.4, the regression equation can be used to predict a value of the response variable (y) for a given value of the explanatory variable (x). For example, the regression line we obtained for the ice cream data was

$$y = 348.9 + 5.9x$$

So when the mean temperature is 10°C, the predicted sales will be 348.9 + 5.9 × 10 = 408, i.e., 408,000 litres. However, this is just a prediction based on a particular sample of 30 observations and, as with any other sample estimate, will be subject to sampling error. If we took data from another 30 periods under identical circumstances, we would almost certainly obtain a different set of results, leading to a similar (but not identical) regression line and hence a slightly different prediction. The key issue is how big the margin of error associated with a regression-based prediction is.

Before we can tackle this issue, we need to be rather more precise about the nature of the prediction we are making. Remember that the regression line gives the *mean (or expected) value* of the response variable for any value of the explanatory variable. In out example, we are implicitly predicting the *average* ice cream sales on all days when the mean temperature is 10°C to be 408,000 litres. However, the average sales of ice cream at a temperature of 10°C is probably not as important to the sales manager as the likely sales in the next period when the mean temperature is expected to be 10°C. In the absence of any other relevant information, our best prediction of sales in the next period is the average sales for a mean temperature of 10°C, but a much larger error could be associated with this figure than in a prediction of a mean value.

Q21. *Why do you think this is?*

In summary, any regression-based prediction could be for either of the following:

■ The *average* value of y at some value of x
■ A *particular* value of y at some value of x

In either case, the value we predict will be the same: the value of y obtained by substituting the relevant value of x into the regression equation. The difference between the two predictions is that the first one will have a smaller margin of error than the second.

8.9 Margin of Error

Most statistics books contain formulae for the margins of error for the two prediction situations we have just described. These formulae are quite complicated to use (even within Excel, although not with more specialised statistical packages), and we will not go into the details here. Instead, we will use simple approximations and concentrate on the general issues so that you understand what influences the margin of error in regression situations. As in Section 7.6, the margins of error are valid if the residuals are normally distributed.

↳ *Predicting a Mean Value*

As a starting point, recall the way in which we calculated the margin of error for an estimated sample mean in Section 7.6:

$$\text{Margin of error} = \pm t_{n-1} \frac{s}{\sqrt{n}}$$

where t_{n-1} is an appropriate percentage point of the t distribution with $n-1$ degrees of freedom and s is the standard deviation of a sample of n values from the relevant population. Apart from the t-value (which reflects the degree of confidence in the result), the margin of error is directly proportional to the variability of the sample (s) and inversely proportional to the square root of the sample size (n).

With one minor change, this result also applies to a regression-based prediction of the mean value of y for a value of x in the *centre of the data* (to be precise, at the mean value of x). The change required is that s is now the *residual standard error (s_e)*, described in Section 8.6, based on $n-2$ degrees of freedom. Consider again our prediction of the average ice cream sales when the mean temperature is 10°C, 408 (in thousands of litres). A temperature of 10°C is quite close to the mean of the x values (which is actually 9.5), and so a 95% margin of error is approximately

$$\pm t_{n-2} \frac{s_e}{\sqrt{n}} = \pm 2.048 \times \frac{40.3}{\sqrt{30}} = \pm 15.1$$

i.e., ±15,100 litres.

Q22. *How would you interpret this result?*

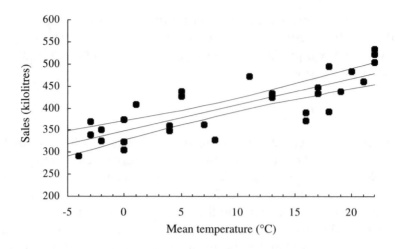

Figure 8.27   Confidence interval for ice cream sales.

So what happens if our prediction is for a value of x not at the centre of the data? As you might anticipate, as we move further out from the centre of the data, our prediction gets more uncertain. This is because any error in estimating the true slope will be amplified the further out we go; that is, the margin of error is greater. It can be shown that if the x values are reasonably uniformly scattered along the x-axis, with no obvious outliers, the margin of error at the extremes of the data will be about twice what it is in the centre.

For example, with the ice cream data, an estimate of the average sales at a mean temperature of 22°C is

$$y = 348.9 + 5.9 \times 22 = 478.7$$

i.e., 478,700 litres. It can be shown that this estimate has a margin of error of 26.0, i.e., 26,000 litres.*

Adding and subtracting the margin of error from the regression line for all values of x gives the error band, or *confidence interval*, around the regression line, as shown in Figure 8.27.

✍ *Predicting a Particular Value*

In this case, we are attempting to predict the y-value of some *particular* point on the scatterplot, which will inevitably be a more uncertain process and lead to a greater margin of error. The main contribution to this margin

---

* The exact margin of error for a given value x is

$$\pm t \times s_e \sqrt{\frac{1}{n} + \frac{(x - \bar{x})^2}{(n-1)s_x^2}}$$

where s_x is the standard deviation of the x values.

of error comes from the natural scatter of the points around the regression line, which is measured by the residual standard error, s_e. As when predicting a mean value of y, the margin of error increases as we go out from the centre of the data, but now by a much smaller amount. In this case, the margin of error might only increase by 5% or 10% as we get to the extremes of the data. This is demonstrated in Figure 8.28, which shows the 95% error band, or *prediction interval*, for a particular period's sales of ice cream for any temperature level. If you look closely at Figure 8.28, you can see that the error limits are slightly curved, but much less so than the corresponding band in Figure 8.27.

If we make the same two predictions as before for sales at $x = 10$ and $x = 22$, the margins of error are 84.0 and 86.6, respectively. So the margin of error at the edge of the data is only 3% greater than in the centre. Furthermore, the limits of error become more parallel and the margin of error more constant as the size of the sample increases.

In summary, if the sample of data is reasonably large ($n > 10$), the margin of error in predicting a particular value of y is more or less constant and is approximately given by

$$\text{Margin of error} = \pm t_{n-2} s_e{}^*$$

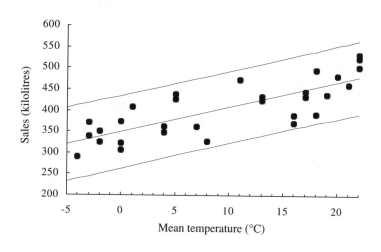

Figure 8.28 Prediction interval for ice cream sales.

---

* The exact margin of error for a given value x is

$$\pm t \times s_e \sqrt{1 + \frac{1}{n} + \frac{(x - \bar{x})^2}{(n-1)s_x^2}}$$

where s_x is the standard deviation of the x-values.

For the ice cream data, this would be $\pm 2.048 \times 40.3 = \pm 82.6$, i.e., $\pm 82,600$ litres.

Q23. *How would you interpret this figure?*

It should be noted that, whether we are predicting a mean or a particular value, predictions made for values of the explanatory variables that are outside the range of the data may not be valid. This is because we have no data with which to judge whether the fitted model is appropriate or not.

8.10   Case Study: Wine Production

Alison Moire, an economist from the Wine Producers Board (WPB), is looking at the relationship between the amount of wine produced in the home market and the average price paid to grape growers. She obtains the data given in Figure 8.29 on the production of chardonnay wine (in thousands of litres) and the average price paid to growers of chardonnay grapes (in thousands of dollars per hectare) for the period 1979–2003. The WPB expects production to increase in 2004 to about 1,650,000 litres of chardonnay and wishes to advise growers of the average price they are likely to receive. A scatterplot of these data is given in Figure 8.30.

The scatterplot indicates that we have a moderately strong negative linear relationship between price and production. That is, the more wine that is produced, the lower the price received by growers. Alison goes ahead and fits a line to the data using Excel. Part of the output is shown in Figure 8.31. The fitted regression equation is therefore

$$\text{Price} = 17.10 - 0.00766 \times \text{production}$$

Since $R^2 = 0.496$, almost 50% of the variation in prices is explained by the linear regression model.

Alison needs to check whether an appropriate model has been fitted, so she plots the residuals against the predicted prices obtained from the regression output, to produce Figure 8.32.

Q24. *Is this an appropriate model for this data set? Are there any outliers?*

Year	Average price (thousands of dollars per hectare)	Production (kilolitres)	Year	Average price (thousands of dollars per hectare)	Production (kilolitres)
1979	1.65	2110	1992	2.45	1643
1980	3.15	1756	1993	3.15	1740
1981	4.60	1365	1994	2.85	1915
1982	6.55	1550	1995	4.45	1870
1983	8.50	1115	1996	4.90	1555
1984	6.35	1600	1997	5.20	1760
1985	3.15	1450	1998	5.25	1525
1986	1.30	1650	1999	4.80	1580
1987	3.35	1510	2000	3.35	1750
1988	8.95	1210	2001	6.20	1500
1989	4.70	1660	2002	11.20	1315
1990	2.60	1800	2003	7.35	1595
1991	3.80	1625			

Figure 8.29   Chardonnay production and average price 1979–2003.

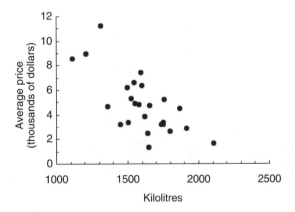

Figure 8.30   Chardonnay production scatterplot.

Regression Statistics			*Coefficients*
Multiple R	0.7040	Intercept	17.105
R Square	0.4956	Production	-0.00766
Standard Error	1.7274		
Observations	25		

Figure 8.31   Selected regression output for WPB data.

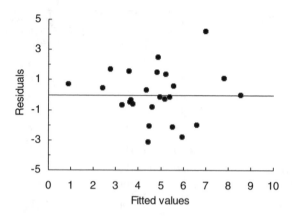

Figure 8.32   Residual plot for WPB data.

Using the regression line, Alison can now predict the average price growers will receive in 2004 to be

$$\text{Price} = 17.10 - 0.00766 \times 1650 = 4.46$$

that is, \$4460 per hectare. Since average production in the period 1979–2003 is 1,606,320, this predicted value has a 95% margin of error of approximately $\pm t_{n-2} \times s_e = \pm 2.069 \times 1.73 = \pm 3.58$. Hence, we are 95% confident that the average price next year will be \$4460 ± \$3580. We should not be surprised that the prediction is not very precise for a particular year because the linear regression only explains about 50% of the variation in average prices.

> ***Q25.*** *Suppose Alison was interested in predicting the average price for all the years when the supply of chardonnay was 1.65 million litres, what would be the 95% margin of error for this prediction?*

8.11   Chapter Summary

In this chapter we have seen how we can use a linear relationship between two variables for prediction. Specifically:

- Always draw a scatterplot of your data to determine whether it is appropriate to fit a straight line to the data.
- A regression line is fitted using the least squares criterion.
- R^2 is used to measure the strength of the linear relationship. This quantity is the square of the correlation coefficient.
- A residual plot gives information about the appropriateness of the model and whether the assumption of a constant standard deviation is likely to be true.

■ Unusual observations can often be identified from the scatterplot of the data or from a residual plot.

■ The fitted regression line can be used for prediction, but it can be misleading to predict the response variable for values of the explanatory variable that are outside the range of the data.

■ A margin of error can be calculated for predictions of either the average value or a particular value of the response variable for a given value of the explanatory variable.

8.12 Exercises

1. The personnel manager conducted an aptitude test on all sales people as they were hired. After a year's employment with the company, the manager obtained each person's total sales figures for the year. The aptitude scores and sales figures (in thousands) are given in Figure 8.33.

   a. Construct a scatterplot of the data, with the response variable plotted on the vertical axis (y-axis).

   b. In what way might the personnel manager use the information in Figure 8.33?

   c. Use Excel to fit a regression of sales on aptitude score. (Save the data for use in Exercise 4.)

   d. Draw your regression line on the scatterplot.

   e. Predict the average sales for employees with aptitude scores of 40, 65, and 90.

2. Artua Smith, a manager of a car rental firm, is trying to estimate the yearly cost of operating rental cars. He has worked out the annual costs of running a small car together with the distance travelled by each car during the year. The graph in Figure 8.34 gives the scatterplot of his data. Artua has produced the output in Figure 8.35 from Excel.

   a. Write down the equation of the regression line.

   b. Explain the meaning of R *Square* in this regression.

Score	Sales	Score	Sales
53	44	81	72
69	63	52	45
62	55	68	59
39	41	78	75
33	30	59	58
102	102	43	40
92	83	46	40
63	56	78	83
78	75		

Figure 8.33   Aptitude scores and first-year sales.

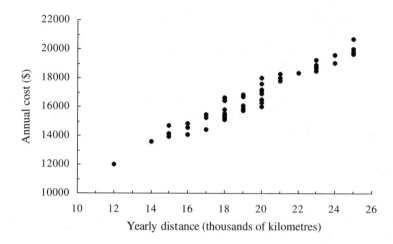

Figure 8.34   Annual cost of hire cars.

    c. Artua has to estimate the cost of running a car that will travel 25,000 km
        during the year. What should he predict as the annual cost of this car?
        What 99% margin of error could he expect from this prediction?
3. In the study of noise levels at London Gatwick Airport described in
    Section 8.1, the scatterplot for aircraft with a steady takeoff that was given
    in Figure 8.11 is reproduced here as Figure 8.36. The aim is to predict the
    perceived noise level (PNL) from the slant distance (SD). The fitted regres-
    sion equation is

$$PNL = 124.3 - 0.024 \times SD$$

    a. Plot the regression line on the scatterplot.
    b. What would be the predicted PNL value for a slant distance of 1000
        metres?
    c. What are the dangers involved in predicting beyond the range of the
        data? Compare your predicted value with the data given in Figure 8.9.
    d. What other variables might be used to improve the prediction of PNL?
4. Consider again the aptitude and sales data given in Exercise 1. A plot of the
    residuals against the fitted values for the regression model is given in Figure 8.37.

Regression Statistics			Coefficients
Multiple R	0.971603955	Intercept	4877.867818
R Square	0.944014246	X Variable 1	607.4185099
Standard Error	460.2415831		
Observations	50		

Figure 8.35   Excel regression output.

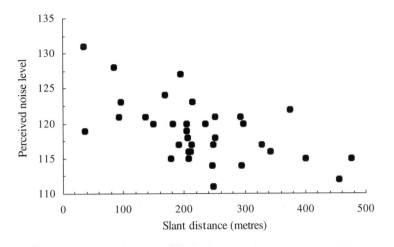

Figure 8.36   Gatwick Airport data — steady takeoff scatterplot.

    a. What conclusions do you draw from this residual plot?

    b. Use Excel to calculate the standard error.

    c. Calculate the 95% margin of error for the average predicted sales for employees with aptitude scores of 40, 65, and 90. Comment on your results.

    d. Suppose that a prospective new employee takes the test and obtains a score of 94. Based on the regression line, give a range of values within which you would expect her sales to be after one year. What assumptions have you made?

5. The scatterplot and the residual plot for a regression of y on x are given in Figure 8.38. One point stands out in each diagram. What do you conclude from these diagrams about the effect of this point on the regression of y on x?

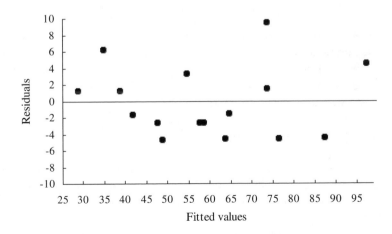

Figure 8.37   Residual plot of sales and aptitude data.

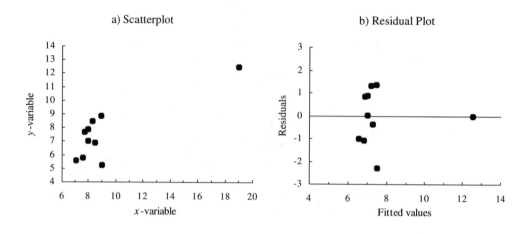

Figure 8.38   Scatterplot and residual plot.

6. From G.E.J. Llewellyn and R.M. Witcomb (in a letter to the editor of *The London Times,* April 6, 1977):

    Sir,

    Professor Mills today (April 4) uses correlation analysis in your columns to attempt to resolve the theoretical dispute over the cause(s) of inflation. He cites a correlation coefficient of 0.848 between the rate of inflation and the rate of change of "excess" money supply two years before.

    We were rather puzzled by this for we have always believed that it was Scottish dysentery that kept prices down (with a one-year lag, of course). To reassure ourselves, we calculated the correlation between the sets of figures shown here [see Figure 8.39].

    We have to inform you that the correlation is −0.868 (which is slightly more significant than that obtained by Professor Mills). Professor Mills says that "Until …a fallacy in the figures [can be shown], I think, Mr. Rees-Mog [editor of the Times], I have fully established my point." By the same argument, so have we.

    Comment on this letter. What conclusions do you draw?

7. The data in Figure 8.40 give the cigarette consumption per person per year in 1930 and the male lung cancer death rate per million males in 1950 for 11 different countries. The cancer data are from 20 years after the smoking data to allow time for cancer to develop.
    a. Draw a scatterplot of these data. Which is the response variable?
    b. Describe the relationship between cancer rate and cigarette consumption. Which country fits in least well with this relationship?

Year	Cases of dysentery in Scotland (in thousands)*	Increases in prices one year later	Year
1966	4.3	2.5	1967
1967	4.5	4.7	1968
1968	3.7	5.4	1969
1969	5.3	6.4	1970
1970	3.0	9.4	1971
1971	4.1	7.1	1972
1972	3.2	9.2	1973
1973	1.6	16.1	1974
1974	1.5	24.2	1975

* Annual Abstract of Statistics, 1976, Table 68.

Figure 8.39   Cases of dysentery in Scotland against price increases.

The regression equation for these data is

$$\text{Cancer rate} = 65.75 + 0.23 \times \text{Cigarette consumption}$$

c. Draw the regression line on your scatterplot. How (if at all) would the slope and intercept change if the outlier identified in (b) were omitted?
d. What is the predicted cancer rate and the corresponding residual for Great Britain?
e. How would you check the assumptions underlying this analysis?
f. The correlation coefficient for these data is 0.741. Explain carefully the meaning of this value. How would this value change if the outlier identified in (b) were omitted?

Country	Cigarette consumption	Lung cancer death rate
Austria	455	170
Canada	510	150
Denmark	380	165
Finland	1115	350
Great Britain	1145	465
Holland	460	245
Iceland	220	58
Norway	250	90
Sweden	310	115
Switzerland	530	250
USA	1280	190

Figure 8.40   Cigarette consumption and lung cancer death rate.

Trip	1	2	3	4	5	6	7	8	9	10
Distance (km)	156	191	356	328	142	174	297	358	167	135
Time (hours)	7.0	5.8	12.1	13.8	6.6	8.9	7.4	7.3	6.3	4.4

Figure 8.41   Distance travelled and time taken.

     g. Is there any evidence from these data that cigarette consumption causes lung cancer? Give reasons for your answer.

8. A homebuilders' association lobbying for various home subsidy programmes argued that, during periods of high interest rates, the number of building permits issued decreased drastically, which in turn reduced the availability of new housing. The data given in the Excel spreadsheet in Figure 3.23 and the scatterplot in Figure 3.24 were presented as part of their argument.

     a. Obtain the linear regression equation, and predict the number of new housing starts given a 10% interest rate.

     b. What do you conclude from your analysis? What would you do next?

9. A trucking company makes regular deliveries of goods to customers in the north of England. Over the last 6 months, trips have been taking increasingly longer than anticipated, sometimes leading drivers to exceed the daily maximum allowable driving time. Data on 10 trips made last week have been recorded and are given in Figure 8.41.

     a. Which are the response and explanatory variables?

     b. Draw a scatterplot of the data.

     c. Describe the relationship between distance travelled and time taken. Comment on any obvious features of your scatterplot.
The regression line for these data is

$$\text{Time} = 3.08 + 0.021 \times \text{Distance}$$

     d. Draw this line on your scatterplot and explain carefully the meaning of the two regression coefficients.

     e. How long do you estimate it will take to make a trip of 420 km? What can you say about the accuracy of this prediction?

     f. The correlation coefficient for the data is 0.677. Explain carefully the meaning of this value, and give two possible reasons why the correlation is relatively low.

10. In seeking to determine just how influential advertising is, the management of a large retail chain collected data over a 16-week period on sales revenue and advertising expenditure from its stores. A scatterplot of the data is given in Figure 8.42.

     a. Comment on the relationship between the two variables.

     b. A regression analysis of sales on advertising gave the following equation:

$$\text{Sales} = -576 + 181 \times \text{Advertising expenditure}$$

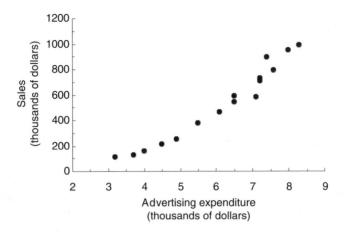

Figure 8.42   Sales against advertising expenditure.

What does the value of the intercept mean? How can it be negative? Make an economic interpretation of the slope.

c.   Given that the residual standard error is 70.95, calculate a 95% confidence interval for the mean sales at the mean level of advertising expenditure of 6.1 (thousand dollars). Interpret the confidence interval.

11.   Workers leave their jobs either to take a different job or to withdraw from the labour force. Economic theory suggests that wages and withdrawal rates are related, and evidence regarding this relationship is important for understanding how labour markets operate. Figure 8.43 lists the withdrawal rates per 100 employees and the average hourly wage in a sample of 15 industries in 2003.

a.   Which variable is the response variable?

b.   Construct a scatterplot of these data.

Industry	Withdrawal Rate	Average Wage ($/hr)	Industry	Withdrawal Rate	Average Wage ($/hr)
1	2.0	8.80	9	1.0	10.59
2	1.8	10.93	10	1.7	9.11
3	1.4	8.83	11	1.7	9.94
4	1.9	6.43	12	3.4	5.37
5	2.3	7.50	13	2.6	6.18
6	3.8	5.54	14	0.7	10.35
7	2.0	7.99	15	1.4	8.20
8	0.5	13.29			

Figure 8.43   Withdrawal rate and hourly wage.

  c. Describe the relationship between the two variables.
     An analysis of these data gave the following regression equation:

$$\text{Withdrawal rate} = 4.86 - 0.35 \times \text{Average wage}$$

  d. Explain how you might use this equation.
  e. Explain what is meant by "fitted values" and "residuals."
  f. Draw a plot of residuals against fitted values. What do you conclude from this plot?
  g. Write a brief summary of the conclusions you draw from your analysis.
12. The marketing manager of Beter Computers wants to determine what affects the company's market share. She believes that the relative price and level of advertising are critical values. Relative price is the ratio of Beter's price to the average for the market as a whole. The manager collects data on the three variables for the past 16 quarters. Explain briefly how she should use these data, and set out the next steps she should take.

Chapter 9

Multiple Regression

9.1 Introduction

Chapter 8 introduced the technique of regression analysis as a means of describing and analysing the relationship between a response variable and an explanatory variable. However, in many real-life situations more than one explanatory variable could be used to predict the response variable. For instance, the salary example in Section 8.1 has two explanatory variables, age and qualification score, both of which could be used to predict salary. Likewise, in the ice cream example, sales may depend not only on daily temperature but also on, for example, the price of the product and the monthly advertising expenditure. Thus we could be looking at a regression analysis that involves up to three explanatory variables. In the moisture content example we could also consider ambient temperature, processing time, and the storage time of the raw material as additional explanatory variables.

Where there are a number of possible explanatory variables, it is inevitable that some will be better predictors than others, and some may not provide us with any additional information in predicting the response variable. Therefore we need to look carefully at which variables we should include in our model and which should be left out. A technique for building regression models involving more than one explanatory variable is called *multiple regression*.

In multiple regression the explanatory variables can take two forms. The variables may be distinctly different, such as age and qualification score, which explain different "dimensions" of variations in salary. Alternatively, one variable may be a function of one or more other variables to represent nonlinearity in a relationship. For example, in the scatterplot of the inflation rate data in Figure 8.6 it is clear that a sensible model of the relationship between interest and inflation rates should take into account the apparent curvature in the data. One way of doing this is to add a second explanatory variable that is the square of the first. Thus we are fitting a *quadratic* (multiple) regression model, and although this now yields two explanatory variables (x and x^2), we still have a two-dimensional relationship. We could even fit a *cubic* model, which would require the addition of an x^3 term, or even a higher-order polynomial. Thus, multiple regression can be used to create both multidimensional linear models and polynomial models in two (or more) dimensions.

As with simple regression, the first step in any multiple regression analysis is to plot the data. For two-dimensional models involving only one explanatory variable this is straightforward. However, with two or more distinctly different explanatory variables, plotting the data is more difficult. A series of two-dimensional scatterplots can be drawn, but sometimes these can be misleading. Computer software is available to examine three-dimensional scatterplots, but most spreadsheet packages do not provide this capability. As mentioned in the previous chapter, this is why residual plots become particularly important in multiple regression.

9.2   Multiple R^2

The key issue in multiple regression is how well the response variable relates to the explanatory variables *collectively* rather than individually. In Section 8.5 we defined R^2 as a measure of the strength of the relationship between the response variable y and the explanatory variable x. More generally, it gives a measure of the relationship between y and a set of explanatory variables. Specifically, $100R^2\%$ of the variation in the response variable is explained by the explanatory variables. The quantity R is called the *multiple correlation coefficient*, but this is generally of limited use in assessing how well a multiple regression model fits the data.

Consider again the salary survey of members of the National Society of Accountants introduced in Section 8.1. To begin with we can calculate the correlation coefficient (r) between the response variable and any explanatory variable and also between any pair of explanatory variables. This gives the *correlation matrix* shown in Figure 9.1, obtained in Excel using the *Correlation* option within the *Data Analysis Tools* menu. Note that the CORREL function in Excel gives the correlation coefficient for only two variables, whereas *Correlation* gives all possible correlations for any number of variables.

From Figure 9.1 we see that the correlation between age and salary (0.601) is much stronger than that between qualification score (*QS*) and salary (0.073). This means that on its own, age explains about 36% of the variations in salary,

	Salary	*Age*	*QS*
Salary	1		
Age	0.6013	1	
QS	0.0728	-0.5261	1

Figure 9.1   Correlation matrix for the salary data.

whereas *QS* explains only about 0.5%. But what if we include both variables? How much of the variation would they jointly explain? One thing we can say is that together they must explain at least 36% of the variation because adding another explanatory variable (*QS*) to the regression against *Age* cannot reduce the amount of explained variability in the response variable. It is tempting to think that together they will explain 36.5% of the salary variation, but the value is much better than this. As we shall see later, $R^2 = 0.571$, so age and *QS* jointly explain about 57% of the variability in salary. Multiple regression often involves a synergy whereby the whole is better than the sum of its parts. The explanation for this is given in the next section.

The reason why R^2 cannot be simply inferred from the correlation matrix is that the explanatory variables are usually correlated with each other. For example, in Figure 9.1, *Age* and *QS* have a correlation of −0.526. In other words, older members tend to have fewer qualifications, so *Age* and *QS* are partly measuring the same thing. Only if the explanatory variables are uncorrelated is it possible to add together their individual R^2 values to get the multiple R^2.

The formula for R^2 is complicated, but most spreadsheets and packages for multiple regression give R^2 (and R) as part of the output, and it is important to understand how to use these values. The use of R^2 is explained in the following sections, but the key issue is how much the value of R^2 increases when another explanatory variable is added. There is no point adding further terms to the model, and thereby increasing its complexity, for very little improvement in the value of R^2. Methods are available to help us decide whether the increase in R^2 is *significant* or not, but these are beyond the scope of this chapter.

9.3 Multiple Regression

To illustrate the process of multiple regression and the use of Excel, we will focus our discussion on the National Society of Accountants salary survey. The Society would like to devise a formula that allows accountants to predict what salary they should be earning. The Society believes a number of explanatory variables could be used in such a formula. The two for which data are available in Figure 8.7 are the age and qualifications of the person surveyed.

> **Q1.** *List other variables that could be important.*

As a first step we should draw three scatterplots, one for each pair of variables. The results are shown in Figure 9.2, where the first two are plots of the response variable (salary) against each of the two explanatory variables (*Age*

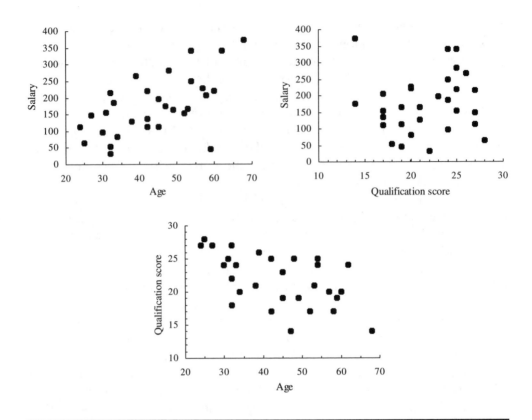

Figure 9.2 Scatterplots for the salary data.

and *QS*). The third plot is a scatterplot of one explanatory variable (*QS*) against the other (*Age*).

> **Q2.** *What can you say about the relationships between the response variable and the two explanatory variables?*
>
> **Q3.** *Which variable would you use to predict salary? Would you use both? Why or why not?*

✎ Regression of Salary against Age and Qualifications

From the scatterplots in Figure 9.2, salary and age appear to be linearly related ($r = 0.601$), but salary and qualifications do not ($r = 0.073$). Suppose we fit a straight-line regression of salary against age. The regression equation is

$$\text{Salary} = -15.5 + 4.36 \times \text{Age}$$

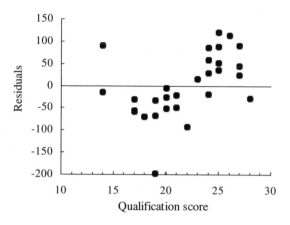

Figure 9.3   Age-related residuals against qualification score.

Q4. How do you interpret the slope (+4.36) in this relationship?
Q5. What should an accountant be earning, on average, at age 45?

Suppose we now calculate the residuals from this line and plot these against the qualification scores to give the scatterplot in Figure 9.3. This scatterplot shows us whether any relationship exists between qualifications and the element of salary that is not age related.

Q6. When we remove the effect of age, do you think a clear relationship exists between salary and qualifications?
Q7. Is it worth including the qualification scores in our regression model? What effect will it have?

Clearly a positive relationship exists between qualifications and the age-related residuals. This tells us that accountants with low qualifications tend to have a salary below what would be predicted for their age, and likewise those with higher qualifications tend to earn more than average for their age. Therefore, a systematic bias exists in the simple model of salary against age that is partly explained by the accountant's qualification score. Despite the low correlation (0.073) between salary and qualifications, there is clear evidence that qualifications are a useful predictor of salary *in conjunction with age*.

This analysis highlights a major drawback in using a simple correlation in a multidimensional situation. The apparently low correlation between salary and qualifications is due to the fact that age is also varying. What we really need is the correlation between salary and qualifications with age held constant. This is called a *partial correlation*. It is beyond the scope of this book to show how to determine partial correlations (it cannot be done easily in Excel), but in this case the partial correlation between salary and qualifications with age held constant is actually 0.573.

✎ Multiple Regression in Excel

In light of the foregoing discussion, we will now fit a regression of salary against both age and qualification score. Again we estimate the intercept and the slopes of the explanatory variables using least squares, which is easily done with Excel. First enter the data on the three variables into an Excel spreadsheet, as in Figure 9.4. (Only the first 21 data values are shown here.) Notice that we have entered variable labels in cells A1:C1, which will subsequently appear in the regression output. Selecting *Regression* in the *Data Analysis Tools*, the regression

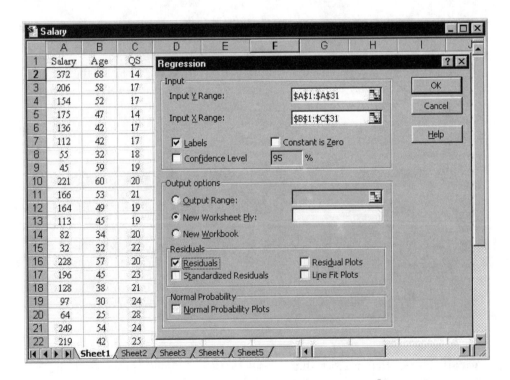

Figure 9.4   Excel spreadsheet windows for fitting a regression model.

SUMMARY OUTPUT			RESIDUAL OUTPUT		
Regression Statistics			*Observation*	*Predicted Salary*	*Residuals*
Multiple R	0.7557		1	239.8	132.2
R Square	0.5711		2	211.4	-5.4
Adjusted R square	0.5393		3	172.9	-18.9
Standard Error	59.789		4	105.2	69.8
Observations	30		5	108.8	27.2
			6	108.8	3.2
	Coefficients		7	56.6	-1.6
Intercept	-362.57		8	241.6	-196.6
Age	6.41		9	259.9	-38.9
QS	11.89		10	226.9	-60.9
			:	:	:

Figure 9.5   Excel output for salary regression.

dialogue window in Figure 9.4 is obtained. Note that in the *Input* boxes we have included the variable names (and checked the *Labels* box) and that the *X Range* box includes both explanatory variables in cells B1:C31. We also check the *Residuals* box to produce a listing of the residuals, which will be used to obtain the residual plot in Figure 9.6.

The relevant parts from the Excel regression output are shown in Figure 9.5; only the first 10 residuals are shown. From this output, the regression equation is

$$\text{Salary} = -362.6 + 6.41 \times \text{Age} + 11.89 \times \text{QS}$$

> *Q8.* *What would you predict an accountant should be earning, on average, at age 45 if she has a qualification score (QS) of only 15?*
>
> *Q9.* *You now have two predictions (Q5 and Q8) of the average earnings of a 45-year-old accountant. For a qualification score of 15, which do you expect to be the more accurate prediction, and why?*

The R^2 value for the regression of salary on age alone is 36.2%. From Figure 9.5, the regression of salary on both age and qualifications gives an R^2 value of 57.1%, an improvement of over 20%. This suggests that both explanatory variables should be used in the regression model, a conclusion that was not obvious from the two-dimensional scatterplots in Figure 9.2.

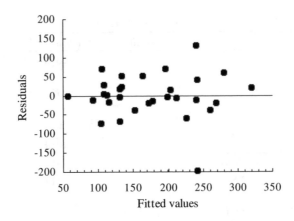

Figure 9.6   Residual plot for salary against age and qualification score.

A plot of the residuals of this regression against its fitted values in Figure 9.6 shows no unusual patterns, although one individual (number 8) seems to have an unusually low salary for his or her age and qualifications.

A margin of error could also be attached to any predicted value of the response variable. The *Standard Error* from Figure 9.5 will be needed for this calculation, which, in multiple regressions, is complicated. It will not be considered further, as specialised statistical packages are needed to give this margin of error.

✍ *Outliers in Regression*

A brief discussion of outliers was given in Section 8.7. The residual plot in Figure 9.6 shows evidence of a possible *outlier* in the data. This individual (number 8) stands out slightly on the first scatterplot in Figure 9.2 but more clearly on the residual plots in Figures 9.3 and 9.6. From the data in Figure 8.7 the actual salary is 45 (thousand dollars), which, from the residual plot in Figure 9.6, is almost 200 (thousand dollars) below what we would predict for a 59-year-old accountant with a qualification score of 19.

> *Q10. What might have caused this unusually large residual of −196.6?*
> *Q11. Omit observation number 8 in the salary data and repeat the regression analysis. In what way has the omission of this point changed the analysis?*

It is important to identify outliers because they can provide potentially misleading results and significantly change the regression model. In the example, the outlier was detected both from the scatterplot of the data and from the residual plot. In this case, omitting the observation does not change the regression equation very much. However, it does reduce the standard error considerably. This in turn reduces the margin of error attached to any predicted value of the response variable.

Outliers are signalled not only by large residuals. In Exercise 5 in Chapter 8, the scatterplot of the data clearly shows the presence of an outlier. However, the residual corresponding to this point is close to 0, as shown in the residual plot. We can see from the scatterplot that deleting this point would result in a substantial change to the regression line. Before deleting any outlier it is important to find out whether there is anything special about that particular observation. Only if something other than its value qualifies an observation as "different" in some way should it be omitted from the data.

Specialised statistical packages will signal points in a regression analysis whose x-values are some distance from the mean of the x-values, or have large residuals, or have a large influence on the regression equation. This feature does not, unfortunately, exist in Excel.

9.4 Polynomial Regression

Consider again the relationship between inflation rates and interest rates for the data given in Figure 8.5. The scatterplot is reproduced in Figure 9.7. We would expect a quadratic curve to give a better fit to these data than a straight line, as can be seen from the best-fitting straight line and curve on the scatterplot. The improvement we get from fitting a quadratic curve can be determined by including an additional variable in our regression to allow for the curvature. For a quadratic model we must add the square of the interest rate variable. Again the best-fitting quadratic model is obtained by the method of least squares using Excel.

✍ *Fitting a Quadratic Model in Excel*

To calculate a quadratic regression model for the inflation rate data, we enter the inflation rates (y) into cells A2:A15 and the interest rates (x) into cells B2:B15,

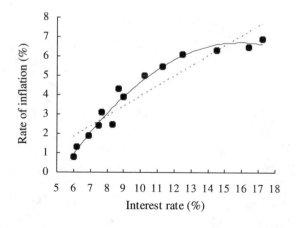

Figure 9.7   Linear and quadratic fits for the inflation data.

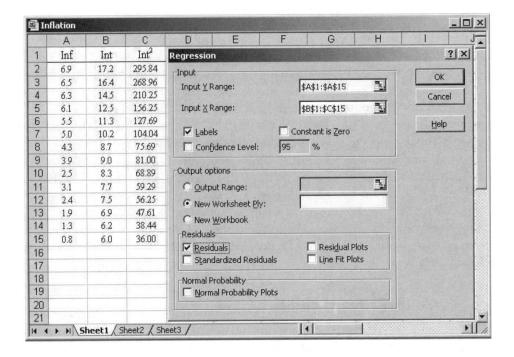

Figure 9.8   Excel spreadsheet windows for fitting a quadratic model.

with the variable names in row 1. We now calculate the x^2 values in cells C2:C15 (i.e., C2=B2*B2, and copy C2 to cells C3:C15) to give the entries in the spreadsheet shown in Figure 9.8. The *Regression* dialogue window is essentially the same as in Figure 9.4. Note that in the *Input X Range* box we have included both the x and x^2 variables. Part of the output from fitting the quadratic model is shown in Figure 9.9, and from this the following quadratic regression model is obtained:

$$\text{Inflation} = -8.23 + 1.88 \times \text{Interest} - 0.06 \times \text{Interest}^2$$

Q12. *Do you think a quadratic model is a sensible one to use? If so, why?*

Q13. *Use this regression equation to predict the rate of inflation given an interest rate of 12%.*

Q14. *What would happen to your predicted inflation rate for a very high interest rate, such as 100%?*

Q15. *Intuitively, what would you expect your model to look like in this last situation?*

SUMMARY OUTPUT

Regression Statistics	
Multiple R	0.9863
R Square	0.9727
Adjusted R square	0.9678
Standard Error	0.3701
Observations	14

	Coefficients
Intercept	-8.2345
Interest	1.8846
Interest^2	-0.0594

RESIDUAL OUTPUT

Observation	Predicted Inflation	Residuals
1	6.61	0.290
2	6.70	-0.199
3	6.60	-0.305
4	6.04	0.057
5	5.48	0.022
6	4.81	0.191
7	3.67	0.634
8	3.92	-0.016
9	3.32	-0.816
10	2.76	0.344
11	2.56	-0.159
12	1.94	-0.042
13	1.17	0.133
14	0.93	-0.135

Figure 9.9   Regression output from the quadratic model.

✥ *Analysis of Residuals*

If we now plot the residuals in Figure 9.9 against the fitted values, we obtain the residual plot in Figure 9.10. This plot shows no evidence of any patterns, indicating that the quadratic model seems to be adequate. Comparing this plot with that in Figure 8.26 shows the improvement brought about by the introduction of the x^2 term into the equation. In Figure 8.26 there was a very obvious inverted U-shaped pattern, indicating nonlinearity in the relationship, which has been removed by the quadratic relationship.

From Figure 9.9, the multiple *R Square* value is 97.3%, which indicates that a quadratic model explains most of the variability in inflation rates. We could

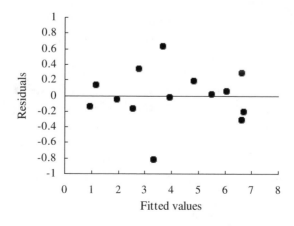

Figure 9.10   Residuals for the quadratic model.

Model	R^2
Linear	0.8700
Quadratic	0.9727
Cubic	0.9742

Figure 9.11 R^2 values for the inflation data.

fit a cubic model, which would require the addition of an x^3 term, or even a higher-order polynomial, but would this be worthwhile? Figure 9.11 gives the R^2-value for three different models.

Q16. What do you conclude from these figures?

♻ *Comparing Alternative Models*

The presence of nonlinearity in data is usually clear from a scatterplot, but this then opens up the question of what form of nonlinear relationship best fits the data. There is no automatic way of deciding this, and in practice it comes down to a process of trial and error. In the inflation rate data we decided to fit a quadratic model largely because it is one of the simplest ways of introducing the required curvature. But other forms of relationship could fit the data equally well and perhaps give more plausible models on the basis of how interest and inflation rates are likely to be related. In essence, we need to consider how the two variables concerned might be related and then try out plausible forms of relationship.

A necessary feature of a polynomial model such as a quadratic or a cubic is that it will have at least one maximum or minimum point. For example, in the quadratic model of the inflation data, inflation will reach a maximum level of 6.7% at an interest rate of about 16%. Thereafter, the model predicts that inflation will fall progressively and ultimately become negative. In this case, it might be more plausible to assume that, as interest rates rise, inflation will also rise, but more gradually. To create a model having this characteristic we may need to transform either the explanatory variable or the response variable or possibly both. For example, we could use the *log* of interest rate as the explanatory variable in a simple regression of y against $\log(x)$. Many other forms of transformation are possible, such as exponential or reciprocal functions, depending on the pattern of curvature shown by the scatterplot. However, to discuss fully the use of such transformations would require a more detailed knowledge of the characteristics of various functions than is appropriate at this level.

9.5 Using Dummy Variables

One of the important aspects of regression analysis is that explanatory and response variables be measurable and give interval data. In a salary survey, variables such as salary, age, qualification score, and years of work experience give interval data and can be used in a regression analysis. However, we could also include attribute data relating to variables such as sex, region, and type of employment, which might be important factors in explaining salary differences. We could assign arbitrary numerical values to the different categories, such as different regions, but the results of the analysis are then dependent on the values assigned. If we assign different numerical values, we will get different answers. One situation when the assignment of numerical values does not matter is when there are only two categories. In this case we can introduce a *dummy variable* that gives each alternative a numerical value (usually 0 or 1) and treat it as if it were interval data. Thus, binary variables such as sex and whether or not a person is in full-time employment can be used as dummy variables in a regression analysis.

The method can be extended to more than two categories. Suppose, for instance, that an attribute variable *region* has three categories, north, central, and south. Two dummy variables would now be needed. The first would have a score of 1 for the north region and 0 for each of the other two regions. The second dummy variable would score 1 for the central region and 0 for the other two. That is, if the two dummy variables are x_1 and x_2, then the north region would take values $x_1 = 1$ and $x_2 = 0$, the central region $x_1 = 0$ and $x_2 = 1$, and the south region $x_1 = 0$ and $x_2 = 0$. This choice for the dummy variables is not unique, but other choices will not affect the results of the multiple regression analysis. For four categories, three dummy variables would be needed, and so on.

Q17. *In the salary survey, what other binary variables could be used in a regression analysis?*

📓 *Moisture Content Example*

Figure 8.3 has 50 pairs of values for the percentage moisture content of an intermediate product and the moisture content of the raw material. The correlation between these two variables was not very high (+0.397), and so the raw material moisture content would not in general be a very accurate predictor of moisture in the intermediate product. However, it subsequently came to light that these moisture contents were measured on two different instruments, the first 25 pairs of values having been taken using instrument A and the last 25 pairs of values recorded using instrument B. A revised scatterplot of the 50 pairs of values showing the instrument used is given in Figure 9.12.

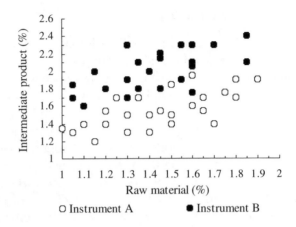

Figure 9.12   Percent moisture content from two instruments.

In a regression analysis, interest would probably centre on predicting the moisture content of the intermediate product (y) given values of both the moisture content of the raw material (x_1) and the instrument used. To include the instrument in the analysis we can introduce a *dummy* variable x_2 that takes the value 0 for one of the instruments, A, say, and the value 1 for the other instrument, B. Using Excel, the fitted regression model is given by

$$y = 0.72 + 0.58x_1 + 0.46x_2$$

> **Q18.** *How would you interpret the three regression coefficients (0.72, 0.58, and 0.46) in this model?*

The effect of introducing the dummy variable x_2 is dramatic. As mentioned earlier, the correlation between the moisture contents of the raw material and the intermediate product was only +0.397, giving an R^2 of 15.7%. Our model, including the x_2 variable, has an R^2 value of 71.7%, an increase of 56% in the explained variability.

Because the dummy variable x_2 only takes value 0 or 1, our regression model corresponds to two distinct parallel straight lines for instruments A ($x_2 = 0$) and B ($x_2 = 1$). That is,

For instrument A: $y = 0.72 + 0.58x_1 + 0.46 \times 0 = 0.72 + 0.58x_1$
For instrument B: $y = 0.72 + 0.58x_1 + 0.46 \times 1 = 1.18 + 0.58x_1$

> **Q19.** *Draw these regression lines on the scatterplot in Figure 9.12, and comment on your analysis.*
>
> **Q20.** *What difference would it make if we were to fit straight lines for instruments A and B individually?*
>
> **Q21.** *How would you decide whether to fit a single model with a dummy variable to represent the instrument or to fit separate models for each instrument?*

9.6 Variable Selection and Model Building

The salary example explored in detail in previous sections is relatively simple in multiple regression terms. The basic data set contains only two explanatory variables, so the decisions to be made are quite straightforward. Given that age is individually the better predictor of salary, it comes down to a question of whether there is anything worthwhile to be gained by also including qualification score. However, with (say) six possible explanatory variables, 63 different combinations of one or more of these variables would need to be considered to find the best model for salary. There is no automatic way of finding the best model in Excel, but this facility is available in specialised statistical packages, such as Minitab. (See http://www.minitab.com.)

We have to consider carefully what we mean by the "best" model. It is not necessarily the model that gives the highest R^2 value, which must be the model that includes all explanatory variables, because, as pointed out earlier, adding an additional variable into a model cannot reduce R^2. In general, we should only include additional variables in a model if they lead to a *worthwhile* increase in R^2. We have to find a good compromise between explaining as much as possible of the variability (i.e., a high R^2) and not having too complicated a model involving unnecessary variables.

The approach most usually adopted is to search for an effective model by proceeding in a "stepwise" fashion, either adding or removing variables one by one. There are two basic strategies that can be used, *forward selection* and *backward elimination*. Forward selection starts with the explanatory variable that has the highest correlation with the response variable and then builds the model up, variable by variable, at each stage including the variable that gives the biggest improvement, providing the improvement is worthwhile. Backwards elimination operates in reverse by starting with a model including all available explanatory variables and removing the ineffective variables one by one, at each stage excluding the variable that has least effect, until all the remaining variables are worthwhile. A more complicated procedure combines the two strategies so that at each stage either an ineffective variable is removed or a worthwhile variable is introduced until no further improvements can be made.

The crucial question is how to judge whether a new variable is worth including or an existing variable ought to be removed. We cannot use R^2 as an indicator because including any variable will necessarily increase R^2 and removing any variable will decrease it. What we really need to know is whether the increase or decrease is *significant*. We could use various criteria to decide whether it is worthwhile moving a variable into or out of the model, but these are generally beyond the scope of this book.

However, a relatively simple backwards elimination procedure exists for finding a "reasonable" model, which makes use of an *adjusted* R^2 you may have noticed in the Excel output in Figures 9.5 and 9.9. Adjusted R^2, which is always less than the actual R^2, takes into account the number of explanatory variables in the model to avoid overestimating the effect on the explained variability of adding a new variable. As a result, it does not necessarily increase with each new variable or decrease with each excluded variable. The procedure goes as follows:

1. Create a regression model using all of the available explanatory variables.
2. Identify the least significant variable (not including the intercept). This is the one with the largest *P-Value*. The *P-Values* are an indication of the contribution of each variable to the model: The smaller the *P-Value*, the greater the contribution. In Figures 9.5 and 9.9 only part of the Excel output was given. However, the full output gives further information about each of the coefficients in the regression model, including their *P-Values*.
3. Exclude this variable from the model and redo the regression analysis. In doing this, you may need first to rearrange the data in your spreadsheet so that the excluded variable is moved to the end and the remaining variables form a continuous block that defines the *X Range*.
4. Observe the effect on the adjusted R^2.
   a. If it increases or remains effectively the same, the excluded variable is not worthwhile and should not be reinstated. Return to step 2 and repeat the process by examining the effect of removing the least significant variable from the new set.
   b. If it decreases, then the explanatory power of the model has fallen and there is a good case to reinstate the excluded variable, particularly if the decrease is "large." All the current variables are therefore worthwhile.

The following case study provides an illustration of this procedure.

9.7 Case Study: Autoparts Ltd.

Autoparts Ltd. is a U.K.-based supplier of automotive parts and spares for European and Asian cars. Its items sell through garages and specialised motor parts retailers throughout the U.K. which is organised into 18 sales regions. Turnover has grown steadily over the last 10 years and last year exceeded £1.3 billion. Though happy with the company's general performance, the sales director believes that more objective regional sales targets ought to be set for the coming

Region	Sales	Mktg	Outlets	Popn	Vehicles	Reps
1	77.51	482	482	2.83	0.83	8
2	58.90	429	262	1.81	0.69	10
3	71.44	84	604	3.67	1.09	11
4	117.45	245	837	5.04	1.60	13
5	98.78	134	819	6.95	2.55	9
6	126.60	461	939	7.43	2.77	14
7	37.99	323	158	1.09	0.39	6
8	78.73	351	497	3.14	0.85	9
9	104.00	298	796	5.59	1.63	14
10	47.32	253	305	2.57	0.95	7
11	61.45	335	394	2.88	0.95	8
12	38.28	329	169	1.32	0.38	7
13	38.62	396	134	1.20	0.42	5
14	61.04	261	330	2.28	0.92	6
15	60.90	474	375	2.08	0.84	9
16	120.93	349	873	5.16	1.81	12
17	75.75	549	422	3.15	1.30	8
18	66.78	286	459	3.12	0.98	10

Figure 9.13   Regional data for Autoparts Ltd.

year, rather than the traditional approach of increasing the previous year's sales by an arbitrary percentage.

Because the regions are all different in terms of sales potential, selling effort, and marketing expenditure, he starts by assembling relevant information regarding regional characteristics and performance, giving rise to the data in Figure 9.13 and the correlation matrix in Figure 9.14. The data show for each region in 2003:

Sales — Regional sales (millions of pounds)
Mktg — Marketing expenditure (thousands of pounds)

	Sales	*Mktg*	*Outlets*	*Popn*	*Vehicles*	*Reps*
Sales	1					
Mktg	-0.0409	1				
Outlets	0.9733	-0.2175	1			
Popn	0.9146	-0.2288	0.9571	1		
Vehicles	0.8729	-0.1495	0.9022	0.9753	1	
Reps	0.8535	-0.0912	0.8585	0.7764	0.6920	1

Figure 9.14   Correlation matrix for Autoparts data.

Outlets — Number of sales outlets selling Autoparts spares
Popn — Population (millions)
Vehicles — Number of registered vehicles (millions)
Reps — Number of Autoparts sales representatives

Q22. *What conclusions do you draw from the correlation matrix in Figure 9.14? Which variables are likely to appear in a regression model of sales?*

From Figure 9.14, it is clear that sales correlate strongly with all but marketing expenditure. It is tempting, therefore, to think that a model for sales should include all variables apart from marketing. However, as we shall see, this turns out to be way off the mark. The four variables that correlate well with sales are also correlated with each other. They are all indicators of regional size in one way or another, and therefore all of them do not need to be present in the model. In contrast, marketing expenditure turns out to be worth including despite its apparently low correlation.

This becomes evident if we fit a regression model to all five explanatory variables, giving the Excel output in Figure 9.15 (note that only relevant parts of the output are shown). From the *P-Value* column in Figure 9.15 it can be seen that only *Outlets* and *Mktg* appear to be making significant contributions. Collectively, all five variables give an adjusted R^2 of 97.25%, but it is likely that some of these variables could be removed without materially reducing R^2.

Regression Statistics				Coefficients	P-Value
Multiple R	0.9903		Intercept	7.80	0.1985
R Square	0.9806		Popn	-5.12	0.3399
Adjusted R Square	0.9725		Vehicles	5.49	0.5969
Standard Error	4.7102		Mktg	0.0395	0.0042
Observations	18		Outlets	0.1332	0.00003
			Reps	-0.1007	0.9156

Figure 9.15   Regression analysis output for Autoparts.

Step	Adjusted R^2	Variables included	Variable coefficient	P-Value	Variable to be removed
1	0.9746	Intercept	7.448	0.12485	
		Popn	-5.228	0.30224	
		Vehicles	5.864	0.53267	←
		Mktg	0.0391	0.00216	
		Outlets	0.1321	0.00000	
2	0.9757	Intercept	6.526	0.14381	
		Popn	-2.386	0.24443	←
		Mktg	0.0417	0.00048	
		Outlets	0.1264	0.00000	
3	0.9749	Intercept	6.220	0.16608	
		Mktg	0.0426	0.00038	←
		Outlets	0.1099	0.00000	
4	0.9440	Intercept	22.581	0.00000	
		Outlets	0.1057	0.00000	

Figure 9.16   Stepwise regression results.

The next stage is to remove variables one by one, starting with *Reps* (since it has the largest *P*-value, 0.9156), to determine the effect on adjusted R^2. Figure 9.16 summarises the results of doing this. Step 1 shows the situation after *Reps* has been removed, where adjusted R^2 has increased to 97.46% and the *P-Value* for *Outlets* and *Mktg* is even lower. At step 2, *Vehicles* is removed, causing adjusted R^2 to increase further to 97.57%. Step 3 removes *Popn,* causing adjusted R^2 to drop slightly, but to no less than its original value. This leaves just the two originally significant variables, *Outlets* and *Mktg*, both of which have a *P-Value* of effectively 0. Finally, removing *Mktg*, as we anticipate, proves to be a step too far, causing adjusted R^2 to fall by about 3%. It appears, therefore, that the most effective model for regional sales involves the number of retail outlets in the region and the level of marketing expenditure:

$$\text{Sales} = 6.22 + 0.043 \times \text{Mktg} + 0.11 \times \text{Outlets}$$

Q23. *The residual standard error for the above model is 4.5. What do you deduce from this value? How accurately does this model predict regional sales?*

The problem of finding a good regression model to fit a large set of data is statistically complicated and beyond what can be covered at an introductory level. In particular, it is possible to formulate more precise rules and procedures for selecting or eliminating variables. Those interested in such methods are advised to consult a more advanced text.

9.8 Chapter Summary

This chapter has extended the methods of simple regression introduced in Chapter 8 to cover situations with more than one explanatory variable or where the relationship involved is nonlinear. Specifically:

- Multiple correlation and multiple R^2 for regression with several explanatory variables.
- Simple correlations and two-dimensional scatterplots can be misleading with more than one explanatory variable.
- Use of derived variables, such as x^2, to model nonlinearity.
- Use of residual plots to identify an inappropriate model and/or outliers.
- Use of dummy variables to represent attribute data in regression.
- Stepwise methods to find an effective model from a large number of explanatory variables.
- Use of adjusted R^2 to avoid overestimating the impact of adding or removing variables.

9.9 Exercises

1. We want to predict the annual salary of a company's executives using months of service and gender as explanatory variables; gender is included so we can compare salaries of female and male executives. The scatterplot in Figure 9.17 is obtained from a random sample of 40 executives from the company. The

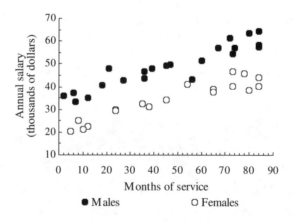

Figure 9.17   Salaries of male and female executives.

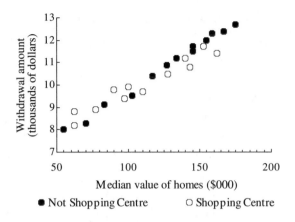

Figure 9.18   Customer withdrawals from ATMs.

following regression equation is obtained, where gender = 1 if the executive is female and gender = 0 if the executive is male:

$$\text{Salary} = 35.1 + 0.3 \times \text{Months} - 14.8 \times \text{Gender}$$

a. What effect do months of service and gender have on annual salary?
b. Predict the annual salary expected by male and female executives after they have been with the company for three years.
c. Plot the regression lines on the scatterplot, and comment on the appropriateness of the model chosen.
d. Explain how you would calculate the residuals from this analysis.

2. A bank would like to be able to predict the total amount of money customers withdraw from automatic teller machines (ATMs) on weekends based on the median value of homes in the neighborhood and on whether the ATM is located in a shopping centre or elsewhere. A random sample of 15 ATMs is selected, with the results shown on the scatterplot in Figure 9.18.

a. Describe the relationship between withdrawal amount and the median value of homes for the two location types.
b. A multiple regression model is

$$y = 6.13 + 0.037\, x_1 - 0.17\, x_2$$

where y is the withdrawal amount, x_1 is the median value of homes, and $x_2 = 1$ for an ATM at a shopping centre and 0 otherwise. Interpret the meaning of the regression coefficients in this equation.
c. Predict the average withdrawal amount for an ATM located in a shopping centre in a neighborhood in which the median home value is $130,000.
d. Would it be appropriate to use the model to predict the average withdrawal amount for a neighborhood in which the median value of homes is $300,000? Why or why not?
e. The R^2 value for this model is 0.96. Explain its meaning.

Manager	Amount of Life Insurance (thousands of dollars)	Annual Income (thousands of dollars)	Marital Status
1	790	60	Married
2	450	85	Single
3	350	75	Single
4	1350	150	Married
5	950	80	Married
6	500	110	Single
7	250	60	Single
8	650	130	Single
9	800	160	Single
10	1400	140	Married
11	1300	120	Married
12	1200	90	Married

Figure 9.19   Life insurance, income, and marital status of managers.

  f. The correlation coefficient between withdrawal amount and median value of homes is 0.978. What conclusions do you draw from your analyses?
3. Figure 9.19 gives the amount of life insurance, annual income for the past year, and marital status of a random sample of 12 managers in the 45–50 age group.
  a. Construct a scatterplot of the amount of life insurance against annual income, with the response variable on the vertical (y) axis.
  b. Describe the relationship between the two variables.
  c. Label each point on the scatterplot with the marital status of the manager. What do you conclude about the impact of marital status on the amount of life insurance purchased?
4. The scatterplot in Figure 9.20 shows the annual expenditure on travel (y) against the family income (x) for a sample of 20 families, some of whom live in rural locations and others of whom live in urban (town/city) locations. Two possible approaches to a regression analysis of the data are being considered:
  i. To separate the data into two groups corresponding to rural and urban location, and to fit a regression line to each group individually.
  ii. To include a dummy variable (d) to represent location (1 = rural, 0 = urban) and determine one multiple regression equation relating y to both x and d.
The regression equations obtained from both approaches are
    Model 1 $y = 7.54 + 0.178\, x$ Rural location
              $y = 4.76 + 0.063\, x$ Urban location
    Model 2 $y = 3.60 + 0.083\, x + 7.86d$

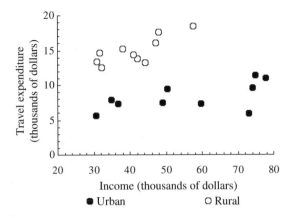

Figure 9.20   Travel expenditure against income.

    a. In the context of this question, explain the meaning of the two regression coefficients in the first equation of model 1 and the three regression coefficients in model 2.
    b. Explain which of the two regression models you would choose and why.
    c. Give two other variables that could help to explain more of the variability in a family's expenditure on travel.
5. Elisa Smith is interested in predicting the sales of road bike tyres. She collects data on two factors she thinks have a major influence on sales: the price of the tyre and the amount spent on advertising. She uses Excel to fit two regression models, the first for sales against price, and the second for sales against both price and advertising. The output for her two models is given in Figure 9.21.

Model 1

Regression Statistics	
Multiple R	0.921
R Square	0.848
Adjusted R Square	0.844
Standard Error	2.608
Observations	40

	Coefficients
Intercept	-264.52
Price ($)	17.05

Model 2

Regression Statistics	
Multiple R	0.922
R Square	0.850
Adjusted R Square	0.842
Standard Error	2.624
Observations	40

	Coefficients
Intercept	-201.62
Price ($)	13.18
Adv (thousands of $)	0.48

Figure 9.21   Regression analyses for road bike tyres.

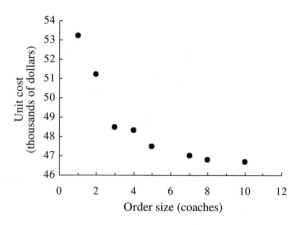

Order Size	Unit Cost
7	47.0
5	47.5
2	51.2
10	46.7
8	46.8
1	53.2
3	48.5
4	48.3

Figure 9.22   Unit cost and order size.

    a. What important step has Elisa missed in her analysis? What sort of consequences could arise from missing this step?

    b. Write down the relationship you would use for predicting tyre sales based on the Excel output in Figure 9.21.

    c. What would be your sales prediction for a tyre priced at $20 with advertising expenditure of $30,000?

    d. To be more confident about your prediction you would need more information about the data set. What extra information do you think will be useful?

    e. What other factors could you consider to improve your prediction?

6. A motor manufacturer produces a small 18-seater coach that is built to order. Due principally to economies of scale from the purchasing of certain key components, the company accountant is aware that unit costs are not constant but decrease with increasing order size. An analysis of the last eight orders completed gives the data and scatterplot in Figure 9.22.

    a. What do you deduce about the form of the relationship between order size (x) and unit cost (y)?

    b. Input the data into an Excel spreadsheet and show that the least squares quadratic model is

$$y = 54.8 - 2.15x + 0.14x^2 \quad (R^2 = 0.950)$$

    What is your prediction of the unit cost of an order for 12 coaches?

    c. An alternative model for y is to use a reciprocal transformation of x, i.e., the explanatory variable $1/x$. This gives the least squares equation

$$y = 46.1 + 7.6/x \quad (R^2 = 0.944)$$

    Now what is your prediction of the unit cost of an order for 12 coaches? Which of the two models (quadratic and reciprocal) do you prefer and why?

Model	Annual sales (in thousands of dollars)	Promotion (in thousands of dollars)	Advertising (in thousands of dollars)	Selling price (in thousands of dollars)
1	6500	23	36	670
2	4700	13	23	570
3	7100	27	35	600
4	4200	28	19	710
5	5100	22	23	690
6	5500	26	25	700
7	7600	35	35	620
8	1800	27	11	730
9	4600	14	23	600
10	2700	26	12	750
11	1300	21	10	680
12	6800	34	36	690
13	4100	13	21	600
14	5400	23	26	640
15	3200	20	10	660

Figure 9.23   Sales data for Hovercut mowers.

7. Hovercut manufactures electric lawn mowers, with 15 different models in the product range. The marketing manager is planning the sales strategy for the coming year and has assembled the data in Figure 9.23 for the current year's sales of each model. Extracts from the Excel regression output for sales in terms of price, advertising, and promotion are given in Figure 9.24.
   a. What is the least squares equation for sales in terms of price, advertising, and promotion? What sales would you predict for a new mower priced at $700 with advertising and promotion budgets of $20,000 and $25,000, respectively?
   b. From the output in Figure 9.24, is it likely that a more effective sales model can be found? Which variable would you eliminate first in a search for a simpler model?
   c. Alternative regression models for sales give the following least squares equations:
      Sales $= 1163.2 + 188.8 \times$ Advertising $- 1.21 \times$ Price (Adjusted $R^2 = 0.916$)
      Sales $= 131.7 + 189.0 \times$ Advertising $+ 9.69 \times$ Promotion
      (Adjusted $R^2 = 0.916$)
      Sales $= 301.7 + 191.5 \times$ Advertising (Adjusted $R^2 = 0.921$)
      Which model would you recommend, and why?
8. Supabytes operates a chain of computer warehouses throughout the United Kingdom. The managing director has requested a profitability analysis of the Leicester store based on data for the last 20 months, which is given in Figure 9.25. It shows the gross operating profit, sales of hardware, sales of software, and the retail price index (RPI) for each month. Extracts from an

Regression Statistics			Coefficients	P-Value
Multiple R	0.9668	Intercept	3071.1	0.2904
R Square	0.9348	Promotion	38.08	0.2998
Adjusted R Square	0.9170	Advertising	170.9	0.00002
Standard Error	544.25	Price	-4.83	0.3010
Observations	15			

Figure 9.24   Excel regression output for Hovercut.

Excel regression analysis of the data are given in Figure 9.26. The objective is to obtain a model for profit in terms of the other three variables.

a. Comment on the goodness of the three variables in explaining profit.
b. Suggest a possible reason for the negative coefficient for hardware sales.
c. How would you proceed with your analysis of the data in light of the output in Figure 9.26?

Month	Profit (in thousands of pounds)	Hardware Sales (in thousands of pounds)	Software Sales (in thousands of pounds)	RPI
1	129	275	197	100
2	134	293	205	101
3	129	275	194	102
4	138	310	209	105
5	137	312	208	108
6	136	295	203	109
7	131	298	202	109
8	131	285	182	110
9	135	308	205	110
10	112	249	157	113
11	130	290	189	115
12	125	275	179	117
13	115	268	162	118
14	132	288	191	118
15	107	205	141	120
16	110	294	167	121
17	128	273	173	121
18	137	305	195	125
19	128	280	180	130
20	121	297	175	131

Figure 9.25   Sales and profit data for Supabytes.

Regression Statistics			Coefficients	P-Value
Multiple R	0.9486	Intercept	-1.881	0.9235
R Square	0.8998	Hardware	-0.129	0.0780
Adjusted R Square	0.8810	Software	0.688	0.00001
Standard Error	3.251	RPI	0.334	0.0193
Observations	20			

Figure 9.26   Excel regression output for Supabytes.

    d. The company has budgeted for hardware and software sales next month of $300,000 and $190,000, respectively. If the RPI stays at 131, what is your estimate of gross operating profit next month?

    e. Months 8 and 17 can be regarded as unusual because month 8 had a greater than usual promotion of a particular high margin software product, and in month 17 a major competitor closed down. Calculate the residuals for these two months, and show whether these effects are evident in the data.

9. The earnings of the chief executive officer (CEO) vary considerably between companies. *Forbes* magazine (May 1995) gave data on the earnings of 20 CEOs, along with details of the person and company in question. It has been suggested that CEO earnings are most influenced by *age, education* level, years *experience* as a CEO, percentage of company *shares* owned, and the company *sales*.

    a. From the extracts of Excel regression output in Figure 9.27, which appear to be the important variables in predicting CEO earnings? Interpret the regression coefficients for these variables.

    b. Are these coefficients plausible? Explain why or why not.

    c. The *education* variable is coded as follows:
      0 = No university/college degree
      1 = Bachelor's degree
      2 = Postgraduate qualification
    Comment on the use of this variable in a regression analysis. How should a variable such as this be included in a regression model?

    d. If the variables *age* and *experience* are omitted from the model, the multiple correlation coefficient becomes 0.897, and all the remaining terms are significant. What does this suggest about the most suitable model for predicting CEO earnings?

Regression Statistics			Coefficient	P-Value
Multiple R	0.9043	Intercept	2672.4	0.0404
R Square	0.8178	Age	-7.758	0.6915
Adjusted R Square	0.7527	Education	-926.5	0.00005
Standard Error	328.59	Experience	9.931	0.3305
Observations	20	Shares	-94.83	0.0204
		Sales	0.072	0.0048

Figure 9.27   Excel regression output for CEO salaries.

Company	P/E Ratio	Profit Margin (%)	Growth Rate (%)
A	11.3	6.5	10
B	10.0	7.0	5
C	9.9	3.9	5
D	9.7	4.3	7
E	10.0	9.8	8
F	11.9	14.7	12
G	16.2	13.9	14
H	21.0	20.3	16
I	13.3	16.9	11
J	15.5	15.2	18
K	18.9	18.7	11
L	14.6	12.8	10
M	16.0	8.7	7
N	8.4	11.9	4
O	10.4	9.8	19
P	14.8	8.1	18
Q	10.1	7.3	6
R	7.0	6.9	6
S	11.8	9.2	6

Figure 9.28   Company financial performance.

10. The data in Figure 9.28 show the price/earnings (P/E) ratio, net profit margin, and annual growth rate for a sample of 19 U.S. companies.
   a. Input the data into an Excel spreadsheet and determine a least squares model to explain the P/E ratio in terms of net profit margin and growth rate.
   b. Which of the two explanatory variables appears to be the best indicator of P/E ratio?
   c. Using Excel, calculate the residuals for your model and draw a scatterplot of residuals against fitted values. Comment on any important features that you can see in the scatterplot.
   d. What would you do next to refine your regression model?
11. Pandora Inc. operates a regional distribution system from each of its seven depots. Every day, the drivers at each depot are allocated a number of customers to whom deliveries have to be made. The driver decides on an appropriate route to visit each of his customers before finally returning to the depot. To assess the effectiveness of the distribution system, the data in Figure 9.29 have been collected from a random sample of 20 trips, showing the time taken, the number of customers visited, the overall size of the load, and the distance covered. Figure 9.30 presents selections from an Excel regression analysis of the data.

Trip	Time taken (hours)	Number of customers	Load (cartons)	Distance (miles)
1	10.5	5	48	250
2	8.2	5	40	158
3	7.2	3	31	156
4	9.7	4	38	162
5	10.9	6	55	200
6	5.7	3	31	131
7	10.1	5	47	196
8	12.8	7	61	232
9	6.4	3	31	138
10	12.8	7	60	238
11	8.5	4	40	158
12	7.1	4	36	139
13	9.2	5	47	178
14	5.3	2	31	126
15	6.9	3	37	139
16	10.3	6	50	201
17	12.7	7	58	227
18	7.5	4	37	145
19	7.7	3	34	159
20	10.6	6	55	189

Figure 9.29   Trip data for Pandora.

a.   Comment on the appropriateness of the number of customers, the load carried, and the distance travelled as an explanation of the time taken. Which is the most significant of these three variables?

Regression Statistics	
Multiple R	0.9744
R Square	0.9494
Adjusted R Square	0.9399
Standard Error	0.5674
Observations	20

	Coefficient	P-Value
Intercept	-0.1206	0.8831
Customers	0.5386	0.1202
Load	0.0613	0.2717
Distance	0.0227	0.0074

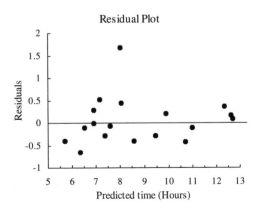

Figure 9.30   Excel regression output for Pandora.

b. How long should it take to make a trip of 210 miles, visiting 5 customers, and carrying a load of 46 cartons?

c. Based on the residual plot in Figure 9.30, is there any evidence that you should investigate further the circumstances of any trip? If so, which one?

d. The correlation coefficient between time taken and distance travelled is 0.928, and the R^2 value for a model with both distance and number of customers is 0.945. Comment on these values regarding the "most appropriate" model for the time taken.

e. What are the potential benefits of a model such as this for a distribution manager?

Chapter 10

Forecasting

10.1 Introduction

As we have seen, one of the main reasons decision making is seldom straight-forward is that most situations involve uncertainty, which makes it difficult to anticipate future events. Indeed, many aspects of business planning and decision making require forecasts of future circumstances, and the only thing that can usually be said with confidence is that forecasts will be inaccurate to some extent. This does not mean that forecasting is a waste of time, because a plan of action based on a well-conceived forecast is likely to be better than instinctive reactions to events that occur.

As with most aspects of management, good forecasting results from a combination of technical skills (data analysis) and informed judgement. It would be foolish to completely ignore any previous data, but it would be equally short-sighted not to make use of relevant contextual information that might have a bearing on future events.

As an introduction to some of the issues that arise in forecasting, we shall consider the following three examples.

Electric Fence Insulators Inc.

Figure 10.1 shows a run chart of the monthly sales figures for an electric fence insulator for the period January 2002 to June 2004.

> *Q1. What is the main feature of these data?*
> *Q2. What would you forecast sales to be in July 2004?*

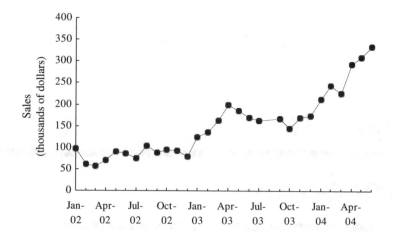

Figure 10.1   Electric fence insulator sales.

📈 Plutomania

Figure 10.2 shows a run chart of the daily attendances at the Plutomania theme park over a 4-week period.

> **Q3.** *What do you notice in particular about these data?*

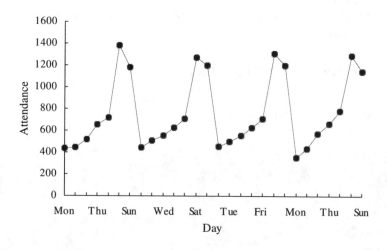

Figure 10.2   Daily attendances at Plutomania.

![] **Pine Products**

Pine Products (Australia) Ltd. manufactures a range of high quality wooden patio furniture that is sold to the domestic market. Every month, the sales director receives a new forecast of quarterly sales for the coming year.

> **Q4.** *Why do you think regular forecasts of quarterly sales would be useful to the sales director?*
>
> **Q5.** *What would be the first thing you would do to produce such forecasts?*

A run chart of quarterly sales (in millions of dollars) from 1995 to 2002 is shown in Figure 10.3.

> **Q6.** *What sort of things can you see from the run chart that might influence your forecasts?*

10.2 Time Series

The data in Figures 10.1, 10.2, and 10.3 are measured at regular time periods. This type of data is called a *time series*. In our examples, we have time series of 30 monthly, 28 daily, and 32 quarterly values, respectively. Our aim is to analyse a time series to detect patterns that will enable us to forecast future values of the series.

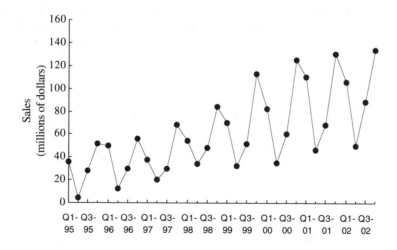

Figure 10.3   Quarterly sales of Pine Products.

The process of forecasting future values of a time series should be, in part at least, an objective one based on an analysis of relevant past values of the series. The process usually involves the following stages:

1. Choose a forecasting model. A forecasting model is a well-defined procedure for calculating a forecast of future values based on the relevant past data. For example, a very simple model might be to average the most recent three values of the series to give a forecast of the next value.
2. Fit the model to these past observations (that is, apply it retrospectively) and obtain the *fitted values* and *residuals*. This tells us what forecasts based on the model would have been in the past (the fitted values) and how inaccurate they were (the residuals).
3. These residuals, as in regression analysis, tell us how well the chosen forecasting model explains the data and provide us with information about its acceptability.
4. If the model is deemed acceptable, we can use it to *forecast* future observations, such as sales in the next quarter.
5. Finally, as new observations become available, we can compare them with our forecasts, leading to an evaluation of *forecasting errors*. This allows us to *monitor* the performance of the model.

When deciding on a suitable model for a time series, it is often useful to identify particular features in the past data so that the series can be broken down (decomposed) into its component parts. The model will specify how these components combine to give the observed values, so we can create forecasts by extrapolating the individual components and then reassembling them according to the specified model. It is rather like forecasting a company's total sales by predicting each product individually and then combining these to give a total sales figure.

Consider the three simple (almost trivial) situations shown in Figure 10.4, where forecasts are to be made for period 7 onward based on the first six values. In all three examples, forecasts of future values would be simple (and highly accurate) because there is a precise pattern in each case. In example A, the past

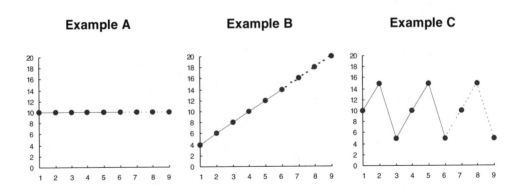

Figure 10.4   Three simple time series.

values are the same, so the *level* of the series is constant or stable and, in the absence of any indications that this situation is about to change, we would forecast a value of 10 in future periods. In example B, the values are clearly not constant, but they are rising by a constant amount per period. In other words there is a stable *trend* upward of 2 each period, so we would forecast values of 16, 18, 20, and so on in future periods. Example C exhibits a regular pattern such that the values rise and fall predictably around a constant underlying average of 10. Again, in the absence of any indications of a change in this pattern, we would forecast future values of 10, 15, 5 repeating every three periods. This kind of regular variation is called a *seasonal* pattern. In a more complex situation, a seasonal pattern may be combined with an upward or downward trend.

In practice, however, forecasting is made more difficult by the presence of a "nuisance element" in most time series: *random variation*. Various unpredictable things almost always occur to disrupt otherwise regular patterns and increase uncertainty. Thus the graphs in Figure 10.4 are more likely to look like those in Figure 10.5. These random disturbances will normally average out to zero in the long run, in which case the uncertainty will not affect the forecasts we make, but it will make them less accurate.

Three standard features (components) of a time series can be identified:

■ *Current level.* The level of the time series in any period is the *expected* value of the variable in that period. This expected value may be constant or may vary from period to period. If the data exhibit no trend, the level is likely to wander up and down while remaining fundamentally horizontal. With a trend, the level will progressively increase or decrease over time. In a series having only a level component (i.e., no trend), forecasts for all future periods will reflect only the current level and so will be the same for all future periods.

■ *Long term trend.* This is a fundamental rise or fall in the data over a sufficiently long time period. In many cases trends are assumed to be linear (and modeled accordingly), but trends can be nonlinear, such as when a product's sales slow down (or even fall) after an initial period of growth.

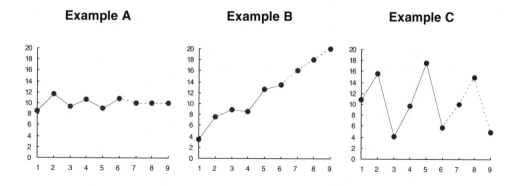

Figure 10.5 Three simple time series with noise.

When a trend is present, forecasts will rise (or fall) from the current level by the amount of the trend per period. For example, if we estimate that the current level of the series is 10 and there is an upward trend of 2 per period, then (in the absence of any other components) forecasts for the following periods will assume that this trend continues, giving values of 12, 14, 16, and so on.

■ *Seasonal pattern.* This arises when the data are influenced by particular effects, such as the weather, that cause a regular and repeating pattern to occur over some period, such as a week or a year. Many things, such as holiday items and sporting goods, have sales patterns that vary predictably through the year. Likewise, the daily sales turnover in a grocery store will often follow an identifiable pattern through the week. When forecasting a series that has a seasonal component, the forecast level (including any trend that might be present) is adjusted by the estimated seasonal pattern.

Any time series has a level and (almost certainly) some random variation, but we need to determine if either of the other two components is also present. If so, we can develop a good forecasting model by decomposing the series into its level (L), a trend (T), a seasonal component (S), and a random noise component. Having estimated L, T, and S from past values of the series, we can then combine them according to our model to forecast future values. By definition, the random component is unpredictable, but it should average out to zero in the long run.

10.3 Estimating the Level

When the data show no trend or seasonality, the current level (L) can be estimated by averaging the observed time series data. If the level is reasonably constant, such as in example A in Figure 10.5, a simple mean of a number of past values is all that is required to estimate the level. Taking the mean of the most recent n values gives an n-point *moving average*, where each of the n values contributes equally to an estimate of the level. In such a situation, the only decision to be made is what portion of the past data is relevant to a future forecast — in other words, the number of values (n) to average. As we saw in Chapter 7, the margin of error for any average is inversely proportional to the square root of the number of values averaged, so with a very stable series the accuracy of moving average forecasts is increased by taking as large a value of n as possible. However, with a very large amount of available data, there will come a point where past values are considered to have little or no relevance to future forecasts.

When the level is changing over time, however, giving equal weight to just the last few observations may not be appropriate. Instead it may be better to give more weight to the most recent values and less and less weight to values further back in time. A forecasting method that does this is called *exponential smoothing*.

✎ *Exponential Smoothing*

Exponential smoothing is an appropriate method of forecasting when the data exhibit no clear trend or seasonality but where the level of the series is less stable. If the level moves up and down from period to period, we need a form

	% Weight for value in period			
	....	t-3	t-2	t-1	t
$\alpha = 0.1$	....	7.3	8.1	9.0	10.0
$\alpha = 0.9$	....	0.1	0.9	9.0	90.0

Figure 10.6   Exponential smoothing weights with low and high values of α.

of average that reacts quickly to these changes in level. One way of achieving this is to use a weighted average where we give more weight to the most recent data and progressively less weight to data further back in time. In particular, an exponentially smoothed forecast gives a specified weight (a proportion between 0 and 1 that is usually denoted by α) to the most recent value, and the weights *damp down* by a factor $1 - \alpha$ for each period that we go back in time. Because the weights are decreasing by a constant factor of $1 - \alpha$, they are decreasing exponentially, hence the name of this type of forecasting.

Unless $\alpha = 1$, because the weights are decreasing exponentially, they never fall to zero, however far back we go, so all previous values in the series contribute to some degree to an exponentially smoothed forecast. The degree to which the previous values contribute depends on the *damping factor* $1 - \alpha$. The smaller the value of α, the larger is the damping factor, and hence the weights diminish more slowly. In contrast, with a high value of α, the weights diminish more quickly, due to a small damping factor. For example, with $\alpha = 0.1$ and $\alpha = 0.9$ the percentage weights for each value in the series going back in time are as shown in Figure 10.6. Notice that a weight is multiplied by the damping factor $1 - \alpha$ to give the weight in the preceding period.

Thus, with a very low value of α, all previous values receive almost the same weight, whereas with a very high value of α, most of the weight attaches to the most recent one or two observations. As a result, a low value of α means that each new value has little impact and the previous level is only slightly modified, resulting in very stable level estimates that take a long time to adjust to sudden changes in the data. In contrast, a high value of α will give much more erratic level estimates that react more quickly to changes in the data. In the extreme, a value of $\alpha = 1$ will put all the weight on the latest value, and all values prior to that are ignored. In practice, however, large values of α lead to erratic forecasts, and for this reason the value of α is usually less than 0.5.

It seems that the calculation of an exponentially smoothed forecast will be difficult on account of the potentially large number of values involved and their different weights. However, it turns out to be quite straightforward using a simple updating procedure whereby forecasts are calculated recursively. Suppose we have made an exponentially smoothed estimate of the level L_{t-1} of the series at time $t-1$. Then when we get a new observation y_t we revise our estimate of the level as

$$L_t = \alpha y_t + (1-\alpha)L_{t-1}$$

Because we are assuming the series shows no trend, forecasts of all future values after time t are given by L_t. That is, the forecast at any future time $t + h$ is given by

$$\text{Forecast:} \quad F_{t+h} = L_t$$

From the smoothing formula we see that $L_1 = \alpha y_1 + (1 - \alpha) L_0$, so we need an estimate of the level at the beginning of the series (L_0) to start the exponential smoothing process. Possible choices for L_0 would be the initial data value, y_1, or the mean of the first few observations. We also need to choose the value of the smoothing constant α, which we consider later.

Insulator Sales

In the insulator sales data in Figure 10.1, the figures are reasonably stable over the first year, followed by a short increasing sequence and then a six-month decreasing sequence. It is possible, therefore, to regard the first two years as showing a fluctuating level or possibly a gradually increasing trend. The last 10 points in the series indicate more clearly a rising trend, but it is not obvious when this trend began and whether the series as a whole should be regarded as consisting of a fluctuating level or a varying trend. If we regard the series as a fluctuating level, we can use exponential smoothing, as described in the previous section.

The exponential smoothing calculations are presented in Figure 10.7, with the insulator sales data given in column B. At this stage we will let $\alpha = 0.2$; in the next section we shall look at how to choose an appropriate value for α. To estimate the initial level (in cell C2) we shall use the first sales value, so $L_0 = 98$. Then, for example, an estimate of the average level in January 2002, when sales are 98 units, will be

$$L_1 = 0.2 * \text{B3} + (1 - 0.2) * \text{C2} = 0.2 \times 98 + (1 - 0.2) \times 98 = 98$$

Because we have used the first actual value to give us our initial level estimate (L_0), the estimated level in month 1 (L_1) is also 98. However, as new and different monthly figures arise, our estimated level in subsequent months will change from month to month. Note also that the first equation uses the cell references (which is what should be entered in the appropriate cell in the spreadsheet) and the second equation uses the actual values.

The estimates of the monthly levels are given in column C in Figure 10.7, with the initial estimate in cell C2. The only calculation required is that of L_1 in cell C3, where cell references are used. This value can then be copied into cells C4:C32. Column D gives our forecasts. Using the forecast formula, we have, for instance, a forecast for sales in March 2002 as

$$F_{\text{Mar-02}} = L_{\text{Feb-02}} = 90.4$$

	A	B	C	D	E	F
1	Month	Sales	Level (L)	Forecast (F)	Error	|Error|
2			98.0			
3	Jan-02	98	98.0	98.0		
4	Feb-02	60	90.4	98.0	-38.0	38.0
5	Mar-02	57	83.7	90.4	-33.4	33.4
6	Apr-02	71	81.2	83.7	-12.7	12.7
7	May-02	90	82.9	81.2	8.8	8.8
8	Jun-02	87	83.8	82.9	4.1	4.1
9	Jul-02	75	82.0	83.8	-8.8	8.8
10	Aug-02	104	86.4	82.0	22.0	22.0
11	Sep-02	89	86.9	86.4	2.6	2.6
12	Oct-02	95	88.5	86.9	8.1	8.1
13	Nov-02	93	89.4	88.5	4.5	4.5
14	Dec-02	80	87.5	89.4	-9.4	9.4
15	Jan-03	124	94.8	87.5	36.5	36.5
16	Feb-03	135	102.9	94.8	40.2	40.2
17	Mar-03	162	114.7	102.9	59.1	59.1
18	Apr-03	200	131.8	114.7	85.3	85.3
19	May-03	185	142.4	131.8	53.2	53.2
20	Jun-03	170	147.9	142.4	27.6	27.6
21	Jul-03	163	150.9	147.9	15.1	15.1
22	Aug-03	167	154.2	150.9	16.1	16.1
23	Sep-03	145	152.3	154.2	-9.2	9.2
24	Oct-03	169	155.7	152.3	16.7	16.7
25	Nov-03	174	159.3	155.7	18.3	18.3
26	Dec-03	212	169.9	159.3	52.7	52.7
27	Jan-04	243	184.5	169.9	73.1	73.1
28	Feb-04	225	192.6	184.5	40.5	40.5
29	Mar-04	294	212.9	192.6	101.4	101.4
30	Apr-04	310	232.3	212.9	97.1	97.1
31	May-04	335	252.8	232.3	102.7	102.7
32	Jun-04	353	272.9	252.8	100.2	100.2
33	Jul-04			272.9		
34	Aug-04			272.9	MAD =	37.8
35	Sep-04			272.9		

Figure 10.7 Exponential smoothing in Excel.

Other forecasts are obtained in a similar way. Since the level estimate for June 2004 is 272.9, this will then be the forecast for July 2004 and beyond.

A plot of the data and the exponentially smoothed forecasts up to September 2004 are given in Figure 10.8. Although the forecasts are increasing steadily, they tend to lag behind the data and do not, therefore, adequately allow for the upward trend in the data, which can also be seen in the increasingly large positive errors at the bottom of column E in Figure 10.7. Additionally, because exponential smoothing is assuming no trend in the data, the forecasts from July 2004 onward do not change, which is possibly unrealistic in light of the steady rise in the data over the preceding 12 months.

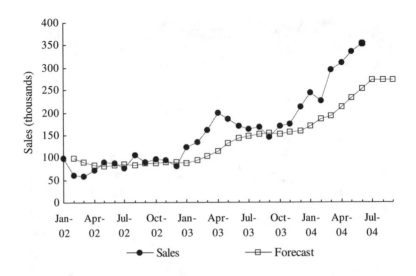

Figure 10.8   Insulator sales data and exponential smoothing forecasts ($\alpha = 0.2$).

⚡ Choosing the Smoothing Constant

The "best" value for the smoothing constant α is found by trial and error to give the smallest overall errors as measured by, for example, the mean absolute deviation (MAD). In Figure 10.7 the errors in column E are obtained by subtracting column D from column B, and the absolute error (i.e., ignoring the sign) is in column F. The formulae in cells E4 and F4 are

$$(E4) = B4 - D4 \quad \text{and} \quad (F4) = ABS(E4)$$

which are copied to the other cells in columns E and F. Notice that we do not include period 1 in the error calculation, for this will always be 0 due to our choice of the starting level L_0 as the first value in the series. Finally, the mean absolute deviation is calculated in cell F34 as the average of cells F4:F32.

For $\alpha = 0.2$ we have, from Figure 10.7, MAD = 37.8. Suppose in our Excel spreadsheet we place the value of α in, say, cell G1 (any cell will do as long as we don't overwrite any other values). Now let the formula in cell C3 involve the cell containing the smoothing constant, as follows:

$$= \$G\$1*B3 + (1-\$G\$1)*C2$$

Note the use of absolute cell reference G1. For an explanation of why these are used, see Section 5 in the *Introduction to Excel* on the CD-ROM. As before, the formula in C3 is then copied into cells C4:C32. Now, to calculate the MAD for any other values of the smoothing constant it will only be necessary to change the value in cell G1. For instance, if 0.8 is entered in cell G1, then we get MAD = 21.1, which is appreciably smaller than the previous value.

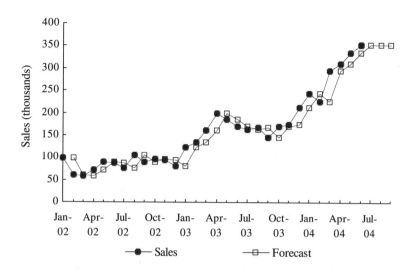

Figure 10.9 Insulator sales data and exponential smoothing forecasts ($\alpha = 1.0$).

By trying other values of the smoothing constant it is relatively easy to home in on the best value. In this case, $\alpha = 1.0$ gives MAD = 20.0, which is the best value obtainable. Using $\alpha = 1.0$ corresponds to taking the most recent value as the forecast and gives the run chart shown in Figure 10.9. It can be seen that these forecasts still lag slightly but are now much closer to the actual sales. More importantly, however, the forecasts for more than one month ahead still show no upward trend.

The need for such a high value of α is due to the upward trend in the latter part of the data. For the forecasts to follow as closely as possible the rise in the data, the highest possible value of α has to be used. This probably suggests that the data should be seen as having a trend rather than a varying level, which will also induce the same upward trend in the forecasts for more than one month ahead. We do this in the following section.

10.4 Forecasting with a Trend

As the preceding example shows, exponential smoothing does not keep up with the data when there is a trend. In such situations we need a method that explicitly models the trend component in the data. A simple way to identify the trend component is to fit a regression model to the time series using the methods of Chapters 8 and 9. Often a linear, or possibly a quadratic, regression will be adequate, where the response variable is the quantity being forecast and the explanatory variable is time. To forecast future values we need to extrapolate the trend, which can be problematic, especially as you try to project further into the future. It is generally necessary to use your judgement for this extrapolation, employing any additional information that is available at the time.

Figure 10.10 shows the linear regression trend line for the insulator sales data. It is clear that the trend has become more markedly upward since September

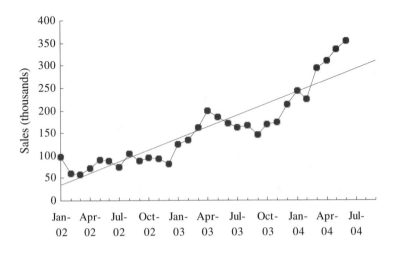

Figure 10.10   Fitting a trend line to the insulator sales data.

2003, and this change is not adequately represented by the linear regression. Little or no upward trend appears prior to December 2002, followed by a steep upward trend over the next five months and then a downward trend for six months. In addition, if there is seasonality present in the data, estimation of the trend by a regression line can be strongly influenced by the pattern of seasonality present. We therefore need a more flexible approach than fitting a straight line or some other simple curve.

A more versatile approach is to apply Holt's *double exponential smoothing*. This method uses the basic exponential smoothing technique not only to determine the current level (L_t) of the series, but also to estimate the current size of the trend per period (T_t). Now when we get a new observation y_t we first revise our estimate of the level as

$$L_t = \alpha y_t + (1 - \alpha)(L_{t-1} + T_{t-1})$$

and then our estimate of the trend as

$$T_t = \beta(L_t - L_{t-1}) + (1 - \beta)T_{t-1}$$

where β is a second smoothing constant.

The first smoothing equation updates the level using a weighted average of the new observation y_t and the level in the previous period (L_{t-1}). However, because of the trend in the data, the previous level (L_{t-1}) will lag the current level and must be adjusted by the estimated trend in the previous period (T_{t-1}). The second smoothing equation then updates the trend, given by the difference between the last two levels.

Forecasts at some future time $t + h$ are given by taking the current level estimate and adding h times the current trend estimate. That is, the forecast is given by

$$\text{Forecast:} \quad F_{t+h} = L_t + hT_t$$

To use double exponential smoothing we need to choose the values of the smoothing constants α and β as well as estimates of the level and trend at the beginning of the series (L_0 and T_0). As before we can use the first data value y_1 to estimate L_0 and any one of a number of possible estimates of T_0, such as:

- No trend, i.e., $T_0 = 0$
- The difference between the first two values, i.e., $T_0 = y_2 - y_1$
- The average difference over a number of early periods

A good choice of starting values gives more realistic forecasts (and hence smaller errors) in the early part of the data, but has progressively less and less effect as time proceeds. If the smoothing constants are not too small, almost any sensible choice of L_0 and T_0 will give more or less the same forecasts after about 15–20 periods.

📇 Insulator Sales

Inspection of Figure 10.1 suggests that a reasonable estimate of the trend at time period 0 is to consider the four consecutive differences between February 2002 and June 2002. The average of these differences is given by one-quarter of the difference between the values in June 2002 and February 2002: $T_0 = (87 - 60)/4 = 6.75$. A simpler approach is to take the difference between the first two values (-38), but this does not give a good estimate of the overall trend in the data, which is clearly positive. In the calculations that follow, we will use $T_0 = 6.75$ in conjunction with the same initial level estimate as before, $L_0 = 98$.

Then, for example, if we set $\alpha = 0.9$ and $\beta = 0.1$, an estimate of the average level and trend in January 2002, when sales are 98 units, will be

$$L_1 = 0.9 \times 98 + (1 - 0.9) \times (98 + 6.75) = 98.68$$

and

$$T_1 = 0.1 \times (98.68 - 98) + (1 - 0.1) \times 6.75 = 6.14$$

In Figure 10.11 the initial estimates of level and trend are in cells C2 and D2 and the calculated values of L_1 and T_1 are in cells C3 and D3, respectively. If cell references are used, then the values in C3 and D3 can be copied into cells C4:D32. Cells E3:E32 give forecasts for one period ahead ($h = 1$). Using the forecast formula, we have, for instance, a forecast for February 2002 as

$$F_{Feb-02} = L_{Jan-02} + T_{Jan-02} = 98.68 + 6.14 = 104.82$$

Other forecasts of the historical data are obtained in a similar way. Future forecasts are obtained from the formula using the June 2004 level and trend values. For instance, a forecast for September 2004 is

$$F_{Sep-04} = L_{Jun-04} + 3 \times T_{Jun-04} = 397.91$$

	A	B	C	D	E	F	G
1	Month	Sales	Level (L)	Trend (T)	Forecast (F)	Error	|Error|
2			98.00	6.75			
3	Jan-02	98	98.68	6.14	104.75	-6.8	6.8
4	Feb-02	60	64.48	2.11	104.82	-44.8	44.8
5	Mar-02	57	57.96	1.25	66.59	-9.6	9.6
6	Apr-02	71	69.82	2.31	59.20	11.8	11.8
7	May-02	90	88.21	3.92	72.13	17.9	17.9
8	Jun-02	87	87.51	3.45	92.13	-5.1	5.1
9	Jul-02	75	76.60	2.02	90.97	-16.0	16.0
10	Aug-02	104	101.46	4.30	78.61	25.4	25.4
11	Sep-02	89	90.68	2.79	105.76	-16.8	16.8
12	Oct-02	95	94.85	2.93	93.47	1.5	1.5
13	Nov-02	93	93.48	2.50	97.78	-4.8	4.8
14	Dec-02	80	81.60	1.06	95.98	-16.0	16.0
15	Jan-03	124	119.87	4.78	82.66	41.3	41.3
16	Feb-03	135	133.96	5.71	124.65	10.4	10.4
17	Mar-03	162	159.77	7.72	139.68	22.3	22.3
18	Apr-03	200	196.75	10.65	167.49	32.5	32.5
19	May-03	185	187.24	8.63	207.40	-22.4	22.4
20	Jun-03	170	172.59	6.31	195.87	-25.9	25.9
21	Jul-03	163	164.59	4.87	178.89	-15.9	15.9
22	Aug-03	167	167.25	4.65	169.46	-2.5	2.5
23	Sep-03	145	147.69	2.23	171.90	-26.9	26.9
24	Oct-03	169	167.09	3.95	149.92	19.1	19.1
25	Nov-03	174	173.70	4.22	171.04	3.0	3.0
26	Dec-03	212	208.59	7.28	177.92	34.1	34.1
27	Jan-04	243	240.29	9.72	215.87	27.1	27.1
28	Feb-04	225	227.50	7.47	250.01	-25.0	25.0
29	Mar-04	294	288.10	12.79	234.97	59.0	59.0
30	Apr-04	310	309.09	13.61	300.88	9.1	9.1
31	May-04	335	333.77	14.71	322.69	12.3	12.3
32	Jun-04	353	352.55	15.12	348.48	4.5	4.5
33	Jul-04				367.67		
34	Aug-04				382.79	MAD =	19.0
35	Sep-04				397.91		

Sheet1 / Sheet2 \ **Sheet3** / Sheet4 / Sh

Figure 10.11   Holt's method for insulator sales data.

A plot of the data and the double exponentially smoothed forecasts are given in Figure 10.12. The plot indicates that double exponential smoothing follows the data much more closely than (single) exponential smoothing. Further, the forecasts of future sales take into account the upward trend in the data.

Q7. Show how the forecast for September 2004 changes if an initial trend estimate of 0 is used.

Q8. What is the effect of using an initial trend estimate of −38?

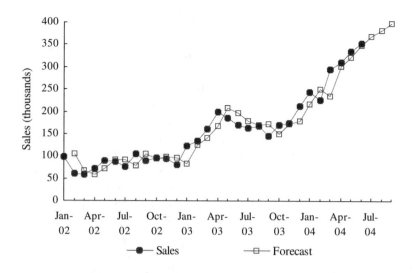

Figure 10.12   Insulator sales data and Holt's method forecasts ($\alpha = 1.0$, $\beta = 0.15$).

To see how well the forecasts fit the data, the mean absolute deviation is calculated, giving MAD = 19.0 for the chosen smoothing constants. As before, if we wish to try other values for the constants, it is more efficient to store them in cells of the spreadsheet and then to reference these cells in the formulae for L and T. For example, suppose that α is in cell K1 and β in cell K2, then the formulae in cells C3 and D3 would be

(C3) = \$K\$1*B3 + (1−\$K\$1)*(C2 + D2) and (D3) = \$K\$2*(C3− C2) + (1− \$K\$2)*D2

By trying other values of the smoothing constants we find that $\alpha = 1.0$ and $\beta = 0.15$ gives MAD = 18.4, so these are probably the best values. This means that we do best by setting the level in any period equal to the observed data value in that period, with the trend relatively unchanged from one period to the next, unless there is a dramatic change in the direction of the data.

Q9.   *Verify the preceding MAD when $\alpha = 1.0$ and $\beta = 0.15$.*

10.5   Including Seasonality

If seasonality is also present in the data, it is important to estimate the size of the seasonal effects so that these can be included in future forecasts. Recall from Section 10.2 that seasonal effects arise when regular patterns occur in the data, such as a weekly, monthly, or annual pattern. For example, in many countries the number of overseas visitors each month follows a predictable pattern, largely because people take vacations at certain times of the year. These patterns will form the basis of the case study on Asian visitors to New Zealand discussed in Section 10.9.

We analyse seasonal effects in two main ways:

- Estimate the size of the seasonal effect and remove the seasonality component from the data. We then have *deseasonalised* data to which we can apply the exponential smoothing methods of earlier sections to obtain a deseasonalised forecast. An actual forecast is then obtained by combining the exponential smoothing forecast with the seasonality component.
- Alternatively, it is possible to include an ongoing deseasonalisation into the basic exponential smoothing process, which deals with seasonality in a way similar to how double exponential smoothing deals with trend. This approach uses a third smoothing constant to update the seasonality estimate period by period and is called Holt–Winters' *triple exponential smoothing*.

In what follows, we shall use the first, somewhat simpler, approach, which assumes that the seasonality component is constant over all the data.

We can estimate the seasonality component using moving averages, provided that the number of periods in the moving average is equal to the number of seasons. Moving averages, briefly referred to in Section 10.3, can be used to give a simple estimate of the level of the series in any period. The size of the seasonal effects can then be estimated by comparing the actual figures with the moving average for the relevant period. This can be done by taking the difference between the actual value and the moving average to give an *additive* measure of the seasonal variation. This is an acceptable way of measuring the seasonal effects if the size of each effect is reasonably constant throughout the data. However, if the run chart indicates that the size of the seasonal effect is related to the size of the data values, it may be more sensible to *divide* the data by the moving average to give a *multiplicative* factor. Alternatively, in this situation the size of the seasonal effects will be additive if the logarithms of the data are analysed, rather than the data themselves.

📉 Seasonality in Plutomania

The Plutomania attendance data shown in Figure 10.2 are now analysed in Figure 10.13 to show how moving averages can be used to estimate additive seasonal factors. In this case additive seasonal factors are appropriate because there is no evidence of a trend in the data and because the size of the seasonal (i.e., daily) variations is reasonably constant from week to week. The objective of the moving average is to remove the seasonal variations so as to expose the underlying level in the data. Since a daily seasonal effect seems present, we use a seven-day moving average so that each day of the week occurs once in every moving average. These calculations are given in Figure 10.13.

The data are given in column C. Column D, headed MA(7), gives the moving averages based on seven days. The Week 1 Thursday result (760.1) is the average (mean) of the attendances on that day (651) and the attendances on the previous

	A	B	C	D	E
	Week	Day	Attendance	MA(7)	Attendance-MA
1	1	Mon	439		
2		Tue	442		
3		Wed	514		
4		Thur	651	760.1	-109.1
5		Fri	717	760.9	-43.9
6		Sat	1379	770.1	608.9
7		Sun	1179	776.0	403.0
8	2	Mon	444	772.3	-328.3
9		Tue	507	771.4	-264.4
10		Wed	555	756.0	-201.0
11		Thur	625	758.6	-133.6
12		Fri	711	759.7	-48.7
13		Sat	1271	759.0	512.0
14		Sun	1197	758.7	438.3
15	3	Mon	452	759.0	-307.0
16		Tue	502	758.9	-256.9
17		Wed	553	764.6	-211.6
18		Thur	627	764.7	-137.7
19		Fri	710	751.4	-41.4
20		Sat	1311	742.0	569.0
21		Sun	1198	744.9	453.1
22	4	Mon	359	750.1	-391.1
23		Tue	436	760.0	-324.0
24		Wed	573	757.4	-184.4
25		Thur	664	749.4	-85.4
26		Fri	779		
27		Sat	1293		
28		Sun	1142		

(Spreadsheet window titled "Plutomania", Sheet1 / Sheet2 / Sheet3)

Figure 10.13   Estimating seasonal factors.

three days (439, 442, and 514) and the following three days (717, 1379, and 1179), that is, the average of seven successive values centred on Thursday. The next value (760.9) is given by the average of that Friday's attendance and the three previous and three following values. The other values in the column are calculated by taking averages of seven successive days in the same way. Moving averages are easily calculated in Excel. For instance, the value in cell D5 in Figure 10.13 is obtained using the formula

$$= \text{AVERAGE(C2:C8)}$$

which is then copied into the other cells in Column D.

The differences between the attendance values and the MA(7) values, given in column E in Figure 10.13, are the additive seasonal factors. Notice that, due to the centring of the moving averages, there are no seasonal factors for the first three and last three days. The seasonal factors are set out again in

	Week 1	Week 2	Week 3	Week 4	Average
Mon		-328.3	-307.0	-391.1	-342
Tues		-264.4	-256.9	-324.0	-282
Wed		-201.0	-211.6	-184.4	-199
Thur	-109.1	-133.6	-137.7	-85.4	-116
Fri	-43.9	-48.7	-41.4		-45
Sat	608.9	512.0	569.0		563
Sun	403.0	438.3	453.1		431

Figure 10.14   Additive seasonal factors.

Figure 10.14, this time grouped by day and week and averaged to give the average seasonal factors in the final column.

> *Q10. What do the averages in the final column of Figure 10.14 tell you?*
> *Q11. What might you expect the mean of the averages in the final column to equal? Check what it is in this case.*
> *Q12. How would you interpret the seasonal factor of –342 for Mondays?*

We can now use the seasonal factors in the final column of Figure 10.14 to deseasonalise the data, where we adjust the actual data values to take out the effect of the particular day of the week. This process is sometimes called *seasonal adjustment*. For example, adding 342 to any Monday figure or subtracting 563 from any Saturday figure will seasonally adjust for that particular day. The deseasonalised data are given in Column D in Figure 10.15.

Exponential smoothing can now be used to forecast values for the deseasonalised data. In this case we use single exponential smoothing, since the data exhibit no trend. The calculations are given in Figure 10.15. Column E gives the forecasts for the level using $\alpha = 0.1$. Column F gives the forecast of the deseasonalised data, and column G gives the forecast for the original data by adding the estimated seasonal factors to the deseasonalised forecasts. For example, the forecast in cell G5 for the first Tuesday is given by adding Tuesday's deseasonalised forecast (781) to Tuesday's seasonal factor (–282). Column H gives the errors (or residuals) from which the MAD of 28.7 is calculated. This is the lowest MAD obtainable and has been found by trial and error by changing the α value,

	A	B	C	D	E	F	G	H	I
					Level	Forecast			
1					(α = 0.1)				
2	Week	Day	Attendance	De-season		De-season	Actual	Error	|Error|
3					781				
4	1	Mon	439	781	781.0				
5		Tue	442	724	775.3	781	499	-57	57
6		Wed	514	713	769.1	775	576	-62	62
7		Thur	651	767	768.9	769	653	-2	2
8		Fri	717	762	768.2	769	724	-7	7
9		Sat	1379	816	773.0	768	1331	48	48
10		Sun	1179	748	770.5	773	1204	-25	25
11	2	Mon	444	786	772.0	770	428	16	16
12		Tue	507	789	773.7	772	490	17	17
13		Wed	555	754	771.7	774	575	-20	20
14		Thur	625	741	768.7	772	656	-31	31
15		Fri	711	756	767.4	769	724	-13	13
16		Sat	1271	708	761.5	767	1330	-59	59
17		Sun	1197	766	761.9	761	1192	5	5
18	3	Mon	452	794	765.1	762	420	32	32
19		Tue	502	784	767.0	765	483	19	19
20		Wed	553	752	765.5	767	568	-15	15
21		Thur	627	743	763.3	766	650	-23	23
22		Fri	710	755	762.4	763	718	-8	8
23		Sat	1311	748	761.0	762	1325	-14	14
24		Sun	1198	767	761.6	761	1192	6	6
25	4	Mon	359	701	755.5	762	420	-61	61
26		Tue	436	718	751.8	756	474	-38	38
27		Wed	573	772	753.8	752	553	20	20
28		Thur	664	780	756.4	754	638	26	26
29		Fri	779	824	763.2	756	711	68	68
30		Sat	1293	730	759.9	763	1326	-33	33
31		Sun	1142	711	755.0	760	1191	-49	49
32	5	Mon				755	413		
33		Tue				755	473	MAD =	28.7
34		Wed				755	556		
35		Thur				755	639		
36		Fri				755	710		
37		Sat				755	1318		
38		Sun				755	1186		

Figure 10.15 Deseasonalisation and forecast for Plutomania.

which is stored in a convenient cell of the spreadsheet. Figure 10.16 shows a run chart of the original data and the one day ahead forecast as well as a forecast for the coming week.

Q13. What are the forecast attendances for Monday and Tuesday in week 6?

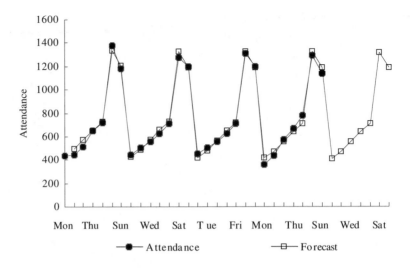

Figure 10.16   Forecasts for Plutomania.

10.6 Residual Analysis

As in regression, the residuals from the fitted model are a useful indication of how well the model fits the data. We could plot the residuals against the forecasts, but because the data are in time order it will be more useful to plot them against time. Any patterns in the residuals over time will then be clear from this plot and, as before, will indicate an inappropriate model. For example, for the Plutomania attendance data the plot of the residuals (given in column H in Figure 10.15) is shown in Figure 10.17.

Q14. What do you conclude from Figure 10.17?

10.7 Forecasting Accuracy

One drawback to using MAD for measuring the residuals in a forecasting model is that it tends to overstate the true accuracy of the model.

Q15. Why do you think this is so?

A more reliable indicator of how good a model is for forecasting is to measure its performance on *new* data, that is, data that have not been used to construct and *calibrate* the model itself. It is a bit like comparing cars that have been

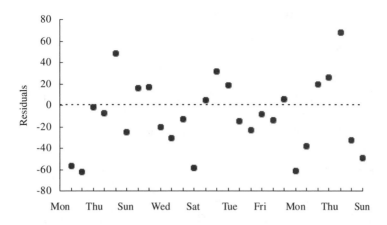

Figure 10.17   Residual plot for Plutomania.

tuned to give optimum performance in particular driving conditions. Arguably, it is better to compare them in more general conditions that were not known in advance rather than in the conditions for which they have been set up.

This implies that we should, if possible, reserve some of our data for testing the model and not use those data in building the model. How much of the data is used for testing and how much for model building will depend largely on the type of forecasting method being used and the amount of data available. When a large amount of data is available, the proportion reserved for testing could be as much as a third. With fewer data available, you may need to use proportionately more for model development in order to produce a reliable model. However, because exponential smoothing is essentially a procedure for short-term forecasting, forecasts would typically be for no more than about 10–20 periods ahead, which will determine the appropriate amount of test data.

It is to be expected that the performance of the model on the test data will be worse, often significantly so, than the *internal* errors obtained on the data used for model development. For example, it is not unusual for the MAD on test data to be as much as 100% greater than on the development data, particularly if a lot of random variation appears in the data or frequent changes occur in the level and/or the trend. After all, if a car has been set up to give good fuel economy under highway conditions, it will not be surprising if it does not perform nearly so well in urban conditions.

This is essentially the same issue as in Chapter 8 when we examined the margin of error of a regression forecast. We saw there that as we extrapolate further out from the centre of the data, the margin of error increases. Using a regression line to predict beyond the range of the data usually leads to greater residuals than predicting within the range of the data. Similarly, using a forecasting model to predict the values of new (or "hold-out") data that were not used to construct the model is likely to be less accurate than for the data on which the model was built.

10.8 Case Study: Pine Products

The quarterly sales of Pine Products from 1995 to 2002 are shown in Figure 10.3. It is clear that both trend and seasonality are present, with higher sales in quarters 1 and 4. In this case, however, the seasonality effect seems to be related to the sales values themselves, for the larger the sales value the greater the seasonal variation. This suggests that multiplicative seasonal factors should be used, which will express the seasonal variation in terms of a proportional variation from the underlying level. Since a trend also exists in the sales data, we will apply Holt's double exponential smoothing to the data after we have removed the seasonality component. Then we will put the seasonality back into the deseasonalised forecast to give forecasts for the original data.

The multiplicative seasonal factors will be estimated based on a four-quarterly moving average. Comparing the data values with the moving averages poses a problem when there is an even number of seasons, because the averages now fall between the middle two of the figures being averaged. For example, the middle of the spring, summer, autumn, and winter quarters is midway between summer and autumn. Before we can work out the seasonal factors, we have to align the moving averages with the data by reaveraging the neighboring MA(4) values using an additional MA(2) average. This is demonstrated in the spreadsheet shown in Figure 10.18.

Notice that the values in cells A2:F33 are positioned at either the top or the bottom of the cell to show the alignment of the moving averages. The important point is that the MA(4) values in column D do not align with the data in column C, but the MA(2) values in column E do. The ratios between columns C and E, given in column F, are used to determine the multiplicative seasonal factors. These ratios are given again in Figure 10.19. The averages in the final column are the multiplicative seasonal factors.

Double exponential smoothing can now be used to obtain forecasts for the deseasonalised data. The calculations are given in Figure 10.20. The deseasonalised data are given in Column D, where, for example, the 1995 Jan–Mar deseasonalised sales (30.2) are given by dividing sales (36) by the Jan–Mar seasonal factor (1.19). The starting values for the level and trend ($L_0 = 30$ and $T_0 = 3$) are chosen from an examination of the first year's *deseasonalised* sales figures. In particular, L_0 is the first value and T_0 is the mean quarterly increase in the first year. The deseasonalised forecasts are given in column G and the forecasts for the original sales, obtained by multiplying the deseasonalised forecasts by the appropriate multiplicative seasonal factors, in column H. Note that MAD = 5.6 was the smallest found by trial and error, based on smoothing constants $\alpha = 0.2$ and $\beta = 0.1$. Finally a plot of the sales data and forecasts is given in Figure 10.21, and a plot of residuals over time is given in Figure 10.22. The forecasts for 2003 take into account the multiplicative seasonal variation.

> *Q16. Do you think you might get a better forecasting model if you set both smoothing constants equal to zero? Why or why not?*

	A	B	C	D	E	F
1	Year	Quarter	Sales	MA(4)	MA(2)	Sales/MA(2)
2	1995	Jan-Mar	36			
3		Apr-Jun	4	30.0		
4		Jul-Sep	28	33.5	31.8	0.882
5		Oct-Dec	52	35.5	34.5	1.507
6	1996	Jan-Mar	50	36.0	35.8	1.399
7		Apr-Jun	12	37.0	36.5	0.329
8		Jul-Sep	30	34.0	35.5	0.845
9		Oct-Dec	56	36.0	35.0	1.600
10	1997	Jan-Mar	38	36.0	36.0	1.056
11		Apr-Jun	20	39.0	37.5	0.533
12		Jul-Sep	30	43.0	41.0	0.732
13		Oct-Dec	68	46.5	44.8	1.520
14	1998	Jan-Mar	54	51.0	48.8	1.108
15		Apr-Jun	34	55.0	53.0	0.642
16		Jul-Sep	48	59.0	57.0	0.842
17		Oct-Dec	84	58.5	58.8	1.430
18	1999	Jan-Mar	70	59.5	59.0	1.186
19		Apr-Jun	32	66.8	63.1	0.507
20		Jul-Sep	52	69.8	68.3	0.762
21		Oct-Dec	113	70.5	70.1	1.611
22	2000	Jan-Mar	82	72.5	71.5	1.147
23		Apr-Jun	35	75.5	74.0	0.473
24		Jul-Sep	60	82.5	79.0	0.759
25		Oct-Dec	125	85.3	83.9	1.490
26	2001	Jan-Mar	110	87.3	86.3	1.275
27		Apr-Jun	46	88.5	87.9	0.523
28		Jul-Sep	68	87.5	88.0	0.773
29		Oct-Dec	130	88.5	88.0	1.477
30	2002	Jan-Mar	106	93.5	91.0	1.165
31		Apr-Jun	50	94.5	94.0	0.532
32		Jul-Sep	88			
33		Oct-Dec	134			

Figure 10.18   Pine Products moving averages.

	1995	1996	1997	1998	1999	2000	2001	2002	Average
Jan-Mar		1.399	1.056	1.108	1.186	1.147	1.275	1.165	1.19
Apr-Jun		0.329	0.533	0.642	0.507	0.473	0.523	0.532	0.51
Jul-Sep	0.882	0.845	0.732	0.842	0.762	0.759	0.773		0.80
Oct-Dec	1.507	1.600	1.520	1.430	1.611	1.490	1.477		1.52

Figure 10.19   Multiplicative seasonal factors.

Pine Products ___ □ X

	A	B	C	D	E	F	G	H	I	J
1					Level	Trend	Forecast			
2	Year	Quarter	Sales	De-seas	(α=0.2)	(β=0.1)	De-seas	Actual	Error	|Error|
3					30	3				
4	1995	Jan-Mar	36	30.2	32.45	2.94	33.0	39.3	-3.3	3.3
5		Apr-Jun	4	7.9	29.90	2.40	35.4	17.9	-13.9	13.9
6		Jul-Sep	28	35.0	32.84	2.45	32.3	25.8	2.2	2.2
7		Oct-Dec	52	34.2	35.08	2.43	35.3	53.6	-1.6	1.6
8	1996	Jan-Mar	50	42.0	38.40	2.52	37.5	44.7	5.3	5.3
9		Apr-Jun	12	23.7	37.48	2.17	40.9	20.7	-8.7	8.7
10		Jul-Sep	30	37.5	39.23	2.13	39.7	31.7	-1.7	1.7
11		Oct-Dec	56	36.9	40.46	2.04	41.4	62.8	-6.8	6.8
12	1997	Jan-Mar	38	31.9	40.39	1.83	42.5	50.6	-12.6	12.6
13		Apr-Jun	20	39.6	41.69	1.78	42.2	21.3	-1.3	1.3
14		Jul-Sep	30	37.5	42.28	1.66	43.5	34.7	-4.7	4.7
15		Oct-Dec	68	44.8	44.10	1.67	43.9	66.8	1.2	1.2
16	1998	Jan-Mar	54	45.3	45.69	1.67	45.8	54.5	-0.5	0.5
17		Apr-Jun	34	67.3	51.34	2.06	47.4	23.9	10.1	10.1
18		Jul-Sep	48	60.1	54.73	2.20	53.4	42.7	5.3	5.3
19		Oct-Dec	84	55.3	56.60	2.16	56.9	86.5	-2.5	2.5
20	1999	Jan-Mar	70	58.8	58.77	2.16	58.8	70.0	0.0	0.0
21		Apr-Jun	32	63.3	61.41	2.21	60.9	30.8	1.2	1.2
22		Jul-Sep	52	65.1	63.91	2.24	63.6	50.8	1.2	1.2
23		Oct-Dec	113	74.4	67.79	2.41	66.1	100.5	12.5	12.5
24	2000	Jan-Mar	82	68.9	69.93	2.38	70.2	83.6	-1.6	1.6
25		Apr-Jun	35	69.2	71.69	2.32	72.3	36.6	-1.6	1.6
26		Jul-Sep	60	75.1	74.22	2.34	74.0	59.2	0.8	0.8
27		Oct-Dec	125	82.3	77.70	2.45	76.6	116.3	8.7	8.7
28	2001	Jan-Mar	110	92.4	82.60	2.70	80.2	95.4	14.6	14.6
29		Apr-Jun	46	91.0	86.44	2.81	85.3	43.1	2.9	2.9
30		Jul-Sep	68	85.1	88.41	2.73	89.2	71.3	-3.3	3.3
31		Oct-Dec	130	85.6	90.02	2.62	91.1	138.5	-8.5	8.5
32	2002	Jan-Mar	106	89.0	91.92	2.54	92.6	110.3	-4.3	4.3
33		Apr-Jun	50	98.9	95.35	2.63	94.5	47.8	2.2	2.2
34		Jul-Sep	88	110.1	100.40	2.87	98.0	78.3	9.7	9.7
35		Oct-Dec	134	88.2	100.26	2.57	103.3	156.9	-22.9	22.9
36	2003	Jan-Mar					102.8	122.5		
37		Apr-Jun					105.4	53.3	MAD =	5.6
38		Jul-Sep					108.0	86.3		
39		Oct-Dec					110.6	168.0		

|◄ ◄ ► ►|\ Sheet1 / Sheet2 \Sheet3 /

Figure 10.20   Deseasonalisation and forecast for Pine Products.

Q17. What do you conclude from Figure 10.22?

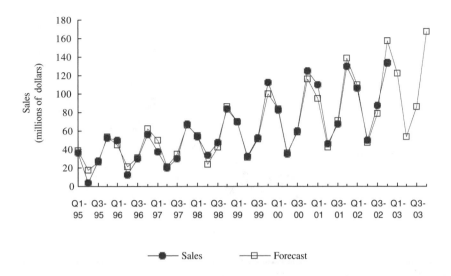

Figure 10.21   Pine Products sales and forecasts ($\alpha = 0.2$, $\beta = 0.3$).

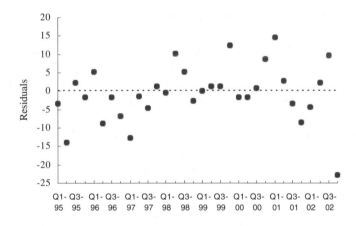

Figure 10.22   Residual plot for Pine Products.

10.9   Case Study: Asian Tourism in New Zealand

The monthly number of Asian visitors to New Zealand from January 1987 through December 1999 is given as a run chart in Figure 10.23. The figures refer primarily to visitors from China, Hong Kong, India, Indonesia, Malaysia, Singapore, South Korea, Taiwan, and Thailand but exclude Japan (*Source*: "International Visitor Arrivals to New Zealand," *Statistics New Zealand*, monthly booklet 1987–2001.).

> *Q18. Accurate forecasts of tourist numbers are important in most countries. Why?*
>
> *Q19. What are the main features of the time series given in Figure 10.23?*

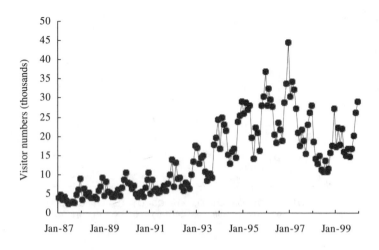

Figure 10.23   Asian visitors to New Zealand, 1987–1999.

A seasonal effect in the tourism numbers clearly exists in Figure 10.23, with numbers highest in the (Southern Hemisphere) summer months. However, the effect increases over time, rather than staying constant. Consequently the multiplicative method is again used to estimate the seasonal factors. Calculations are given in Figure 10.24, where only the data for the first and last 12 months are shown. Since there is a 12-month seasonal effect, MA(12) followed by MA(2) has been used in columns C and D, and the ratio of the data to the MA(2) values is calculated in column E. From the figures in column E the average monthly multiplicative seasonal factors can be calculated; these are given in column F. Finally, the deseasonalised data, given in column G, are obtained by dividing the original data on visitor numbers (column B) by the appropriate seasonal factors.

Q20. How do you interpret the seasonal factors in column F of Figure 10.24? What do you conclude from these values?

The procedure for calculating the monthly forecasts is exactly the same as in Figure 10.20 for Pine Products. In this case we used values of $L_0 = 3800$ and $T_0 = 0$, although with data for 13 years, almost any sensible starting values will converge to similar forecasts by the end of the data. A good choice of smoothing constants is $\alpha = 0.35$ and $\beta = 0.15$, giving MAD = 1409 and the residual plot shown is in Figure 10.25.

	A	B	C	D	E	F	G
1	Month	Visitors	MA(12)	MA(2)	Visitors / MA(2)	Seas Fact	De-seas
2	Jan-87	3,816				1.00	3816
3	Feb-87	4,734				1.25	3787
4	Mar-87	3,492				1.04	3358
5	Apr-87	4,080				1.04	3923
6	May-87	3,042				0.80	3803
7	Jun-87	2,352	4,110			0.68	3459
8	Jul-87	2,616	4,084	4,097	0.639	0.81	3230
9	Aug-87	2,840	4,215	4,150	0.684	0.80	3550
10	Sep-87	2,608	4,332	4,273	0.610	0.67	3893
11	Oct-87	4,752	4,354	4,343	1.094	1.04	4569
12	Nov-87	6,140	4,536	4,445	1.381	1.24	4952
13	Dec-87	8,844	4,664	4,600	1.923	1.58	5597
146	Jan-99	17,200	17,480	17,348	0.991	1.00	17200
147	Feb-99	22,282	17,799	17,640	1.263	1.25	17826
148	Mar-99	17,797	18,223	18,011	0.988	1.04	17113
149	Apr-99	21,931	18,604	18,414	1.191	1.04	21088
150	May-99	16,070	19,340	18,972	0.847	0.80	20088
151	Jun-99	15,026	19,498	19,419	0.774	0.68	22097
152	Jul-99	16758				0.81	20689
153	Aug-99	14573				0.80	18216
154	Sep-99	16692				0.67	24913
155	Oct-99	20244				1.04	19465
156	Nov-99	26264				1.24	21181
157	Dec-99	29136				1.58	18441

Sheet1 / Sheet2 / Sheet3 / Sheet4

Figure 10.24   Asian visitors spreadsheet.

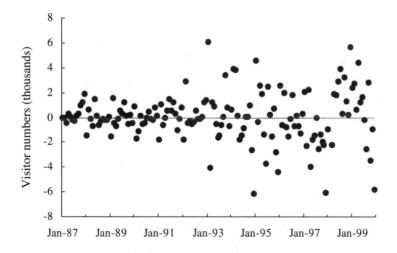

Figure 10.25   Residual plot for Asian visitors.

Q21. What do you conclude from the residual plot in Figure 10.25?

Forecasts of future visitor numbers for 2000 and 2001 are given in Figure 10.26, where the MAD is 2924. As indicated in Section 10.7, the accuracy when forecasting "new" data, in this case up to two years ahead, is appreciably less than that for the data from which the model was constructed. However, despite some large forecasting errors in certain months, for instance, January 2001 and July 2001, the forecasts follow fairly accurately the variations in the actual numbers, caused both by local trends in the data (year by year variation) and seasonal effects (monthly variation).

However, despite the relative accuracy of the forecasts, this example also illustrates some of the difficulties and potential pitfalls involved in trying to forecast the future. What would be a forecast of Asian tourism numbers in five years' time? What forecast would you have made at the end of 1996 for the tourism numbers in February 2000? Could the Asian financial crisis and its effect on tourism numbers have been predicted? Likewise, the terrorist attack in the United States in September 2001 had an impact on air travel worldwide in the immediately succeeding months, although it does not seem to be too evident in Figure 10.26. What effect on travel to and from Asia did the SARS epidemic have in 2002/2003? In general, the further ahead we attempt to forecast, the greater will be the potential impact of major events such as these and, as a consequence, the less accurate such forecasts are likely to be.

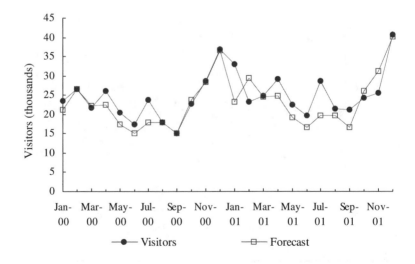

Figure 10.26   Actual and forecast numbers of Asian visitors, 2000–2001.

10.10   Chapter Summary

This chapter has been concerned with the issues that arise in forecasting future events and has discussed some forecasting methods. Specifically:

- The different stages involved in forecasting:
  - choosing a forecasting model
  - fitting and checking the model
  - forecasting future observations and comparing those forecasts with new data
- The three main components of a time series:
  - current level
  - long term trend
  - seasonality
- Exponential smoothing to estimate the current level
- Double exponential smoothing when a trend is present in the data
- Forecasting data with a seasonal component by:
  - estimating the size of the seasonal factors and removing the seasonality component
  - applying exponential smoothing to the deseasonalised data
- How to choose the starting values and smoothing parameters for exponential smoothing
- Using residual analysis to examine the suitability of the model.
- Calibrating the chosen forecasting method on part of the data and using the remaining data to test the accuracy of the fitted model

10.11   Exercises

1. Hickson Tools is trying to forecast sales of its P34 chain saw. A run chart of monthly sales of chain saws is given in Figure 10.27.
   a. Do you think the data show a trend? Can you see any seasonal patterns?
   b. What forecasting methods do you think are appropriate to predict likely sales in January 2005?
   c. Explain how you would go about forecasting sales of the chain saw in January 2005?
2. The sales of videotapes have been recorded over the past three years, and a plot is given in Figure 10.28. The table shown in Figure 10.29 has been produced using exponential smoothing with $\alpha = 0.5$.
   a. What are the main components of this time series?
   b. What are the forecast sales for July 2004?
   c. Comment on how much weight your forecast places on the 2001 data.
   d. How would you judge whether this is a good forecasting method?
3. Mildred Homes, the chief accountant, is drawing up a budget for next year's electricity usage. She has available the electricity usage for the past three years, as shown in Figure 10.30.
   a. What are the most important components of this time series?

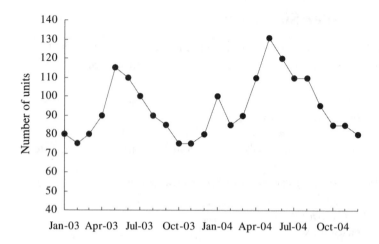

Figure 10.27   Sales of chain saw model P34.

b. Mildred has heard that exponential smoothing is a good forecasting technique. Explain how you would use this approach in this situation.

c. She estimates the total electricity usage next year will be 18,000 kWh. Explain how you would estimate the usage in June 2004. (no calculations are required). Will it be less or more than average monthly usage?

4. Wintago Hospital wishes to forecast weekly patient admissions to determine the number of nurses it needs to employ. The admissions over the last 28 weeks are shown in the run chart in Figure 10.31.

a. Is there any trend in the data? Can you see any seasonal patterns?

b. The hospital admissions manager decides to try exponential smoothing on the data, with a smoothing constant of 0.1, and creates the spreadsheet

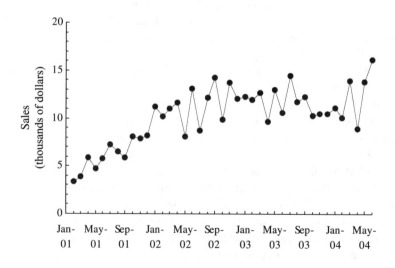

Figure 10.28   Video sales.

	A	B	C	D	E
1	Month	Sales	Level	Forecast	Error
2	Jan-01		3.40		
3	Feb-01	3.4	3.40		
4	Mar-01	3.9	3.65	3.40	0.50
5	Apr-01	5.9	4.78	3.65	2.25
6	May-01	4.7	4.74	4.78	-0.08
7	Jun-01	5.8	5.27	4.74	1.06
8	Jul-01	7.2	6.23	5.27	1.93
9	Aug-01	6.5	6.37	6.23	0.27
10	Sep-01	5.9	6.13	6.37	-0.47
11	Oct-01	8.1	7.12	6.13	1.97
12	Nov-01	7.9	7.51	7.12	0.78
13	Dec-01	8.2	7.85	7.51	0.69
14	Jan-02	:	:	:	:
15	:	:	:	:	:
16	Dec-03	:	:	:	:
17	Jan-04	11.1	10.90	10.69	0.41
18	Feb-04	10.1	10.50	10.90	-0.80
19	Mar-04	13.9	12.20	10.50	3.40
20	Apr-04	8.9	10.55	12.20	-3.30
21	May-04	13.8	12.17	10.55	3.25
22	Jun-04	16.1	14.14	12.17	3.93

Figure 10.29 Extract from Excel spreadsheet for exponential smoothing ($\alpha = 0.5$).

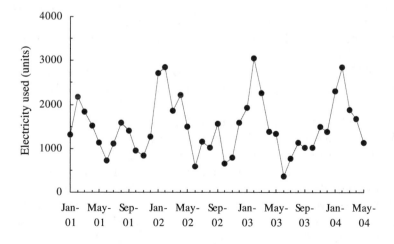

Figure 10.30 Electricity usage.

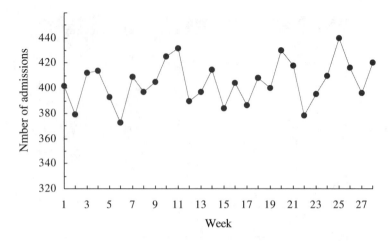

Figure 10.31 Hospital admissions.

shown in Figure 10.32. From this spreadsheet, what would be the forecast number of admissions in week 29? Is this forecast realistic? Why or why not?

   c. What other forecasting method might be appropriate for this situation?

5. A sporting goods retailer has collected quarterly figures for the value of sales (in thousands of dollars) for all its shops in Victoria. A plot of the data is given in Figure 10.33. The retailer wishes to forecast quarterly sales for 2005.

   a. What are the main components of this time series?

   b. What method of forecasting would you advise it to use for its quarterly forecasts?

   c. How do you think you could forecast next year's total sales?

   d. What should you do before using exponential smoothing on these data?

   e. The quarterly seasonal factors for this set of data are given in Figure 10.34. Explain what these quarterly seasonal factors mean.

   f. If the forecast for total sales next year is $3,200,000, estimate the sales for each quarter.

6. Joel Hickman, the manager of Europtour, is looking at forecasting next year's sales of package vacations to Europe. He has collected data on sales over the past four years. A plot of the sales data is given in Figure 10.35. Joel has prepared the table given in Figure 10.36 from a spreadsheet for his analysis.

   a. What forecasting method would you choose?

   b. Forecast the sales for the first quarter of 2005.

   c. How can Joel tell whether an appropriate method of forecasting has been used in this case?

7. The data given in Figure 10.37 show the daily number of patients at the accident and emergency (A&E) unit of the Shepshed Cottage Hospital during the month of September.

   a. Complete the moving averages in Figure 10.37 as far as possible.

	A	B	C
1	Week	Admissions	ES (0.1)
2			402
3	1	402	402.0
4	2	379	399.7
5	3	412	400.9
6	4	414	402.2
7	5	393	401.3
8	6	373	398.5
9	7	409	399.5
10	8	397	399.3
11	9	405	399.9
12	10	425	402.4
13	11	432	405.3
14	12	390	403.8
15	13	397	403.1
16	14	415	404.3
17	15	384	402.3
18	16	404	402.4
19	17	386	400.8
20	18	408	401.5
21	19	400	401.4
22	20	430	404.2
23	21	418	405.6
24	22	378	402.8
25	23	395	402.1
26	24	410	402.9
27	25	440	406.6
28	26	416	407.5
29	27	396	406.4
30	28	420	
31	29		

Figure 10.32 **Exponential smoothing for hospital admissions.**

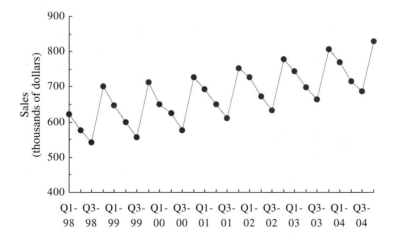

Figure 10.33 **Quarterly sales data.**

Quarter	Seasonal Factor
1	25.1
2	-25.9
3	-71.3
4	72.1

Figure 10.34   Seasonal factors.

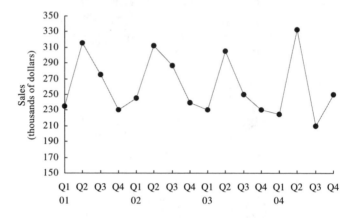

Figure 10.35   European vacation sales 2001–2004.

Quarter	Sales	MA(4)	MA(2)	Sales-MA(2)
01-Q1	235			
Q2	315	263.75		
Q3	275	266.25	265.00	10.00
Q4	230	265.50	265.88	-35.88
02-Q1	245	268.50	267.00	-22.00
Q2	312	271.00	269.75	42.25
Q3	287	267.25	269.13	17.88
Q4	240	265.50	266.38	-26.38
03-Q1	230	256.25	260.88	-30.88
Q2	305	253.75	255.00	50.00
Q3	250	252.50	253.13	-3.13
Q4	230	259.50	256.00	-26.00
04-Q1	225	249.50	254.50	-29.50
Q2	333	254.50	252.00	81.00
Q3	210			
Q4	250			

Figure 10.36   Moving averages for European vacation sales.

Date	Day	Patients	Moving Average
Sep-01	Mon	43	
Sep-02	Tue	47	
Sep-03	Wed	41	
Sep-04	Thu	43	42.43
Sep-05	Fri	45	41.00
Sep-06	Sat	44	39.29
Sep-07	Sun	34	38.71
Sep-08	Mon	33	38.00
Sep-09	Tue	35	37.71
Sep-10	Wed	37	37.29
Sep-11	Thu	38	38.43
Sep-12	Fri	43	38.86
Sep-13	Sat	41	38.14
Sep-14	Sun	42	37.86
Sep-15	Mon	36	37.86
Sep-16	Tue	30	38.14
Sep-17	Wed	35	38.29
Sep-18	Thu	38	38.43
Sep-19	Fri	45	38.14
Sep-20	Sat	42	39.43
Sep-21	Sun	43	
Sep-22	Mon	34	
Sep-23	Tue	39	
Sep-24	Wed	30	
Sep-25	Thu	42	
Sep-26	Fri	45	
Sep-27	Sat	38	
Sep-28	Sun	33	
Sep-29	Mon	38	
Sep-30	Tue	39	

Figure 10.37   Patient numbers at the A&E unit.

Day	Mon	Tue	Wed	Thu	Fri	Sat	Sun
Factor	-4.05	-3.71	-4.10	1.11	5.54		

Figure 10.38   Additive seasonal factors.

Quarter	1	2	3	4	5	6	7	8	9	10	11	12
% incorrect items	17.3	19.2	20.1	18.7	19.5	22.1	22.0	23.8	24.0	24.7	24.5	25.3

Figure 10.39   Items at variance with stock records.

b. To measure the seasonal effects (daily variations) in the data, you decide to calculate additive seasonal factors. Why would you choose additive rather than multiplicative factors?

c. The first five additive seasonal factors are given in Figure 10.38. Calculate the factors for Saturday and Sunday. Explain what the seasonal factors mean, and justify the pattern of values in this case.

d. What are the deseasonalised numbers of patients for September 29 and 30? Would the deseasonalised figures exhibit more or less variability than the moving averages? Explain your answer.

8. A company performs a quarterly stock check in its main warehouse in order to reconcile the values shown on the computerised stock record. Over the last three years, the percentage of items showing a physical count that is at variance with the stock record has been steadily rising, as shown by the data in Figure 10.39.

a. Explain why single exponential smoothing would be an inappropriate method of forecasting the future percentage of incorrect items for these data as they stand.

	A	B	C	D	E
	Year	Month	Sales	MA(4)	MA(2)
1	1999	Jan-Mar	101		
2		Apr-Jun	56	94	
3		Jul-Sep	68	96	95.0
4		Oct-Dec	151	96	96.0
5	2000	Jan-Mar	109	97	96.5
6		Apr-Jun	56	95	96.0
7		Jul-Sep	72	91	93.0
8		Oct-Dec	143	90	90.5
9	2001	Jan-Mar	94		
10		Apr-Jun	52		
11		Jul-Sep	76		
12		Oct-Dec	155		
13	2002	Jan-Mar	105		
14		Apr-Jun	60		
15		Jul-Sep	68		
16		Oct-Dec	143		

Electric Blanket Sales.xls — Sheet1 / Sheet2 / Sheet3

Figure 10.40   Electric blanket sales.

    b. Use an appropriate exponential smoothing method to forecast the percentage of incorrect items in quarter 13.

9. The first three columns in Figure 10.40 give the sales of a model of electric blanket (in thousands of units) over a four-year period.

    a. Give two reasons why it would be inappropriate to use linear regression analysis to forecast sales in 2003.

    b. A moving average analysis of the first two years is given in columns D and E in Figure 10.40. Complete this analysis, and determine the quarterly seasonal factors.

    c. Assuming that no trend exists in the data, provide forecasts of quarterly sales for 2003.

    d. Calculate the mean absolute deviation (MAD) over the period 1999–2002, and hence comment on the likely accuracy of your forecasts for 2003.

Chapter 11

Statistical Process Control

11.1 Introduction

Statistical process control (SPC) is a collection of management and statistical techniques whose objective is to bring a process into a state of stability or control and then to maintain this state. All processes are variable, and being in control is not a natural state. Statistical process control has been found to be an effective way to improve product and service quality. There are many management issues, as well as statistical techniques, that need to be considered to ensure the successful use of statistical process control. Quite often SPC fails in companies, not because of the technicalities, but because of a poor understanding of the important management issues. We shall be more concerned with the concepts and management issues involved with SPC than with the technicalities.

Statistical process control is part of an overall plan to reduce variation and improve business processes. In *The Team Handbook* (Oriel Consulting, Madison, WI, 1988), Scholtes outlines a blueprint for improvement, which he calls the "five stage improvement plan." A schematic of the plan is shown in Figure 11.1.

Statistical process control plays a part in reducing variation and in planning for continuous improvement, the last two stages in the improvement plan. Many companies are so involved in firefighting problems that they never even get to understand their processes. Other companies that are committed to process improvement (often in the manufacturing sector) are mainly concerned with trying to eliminate errors and removing slack from their systems. Companies that apply SPC and plan for improvement do enjoy substantial benefits from doing so. For example, a packaging company improved the quality of the corrugated board it manufactured by applying statistical process control. First they collected data on the thickness of the board every half-hour and plotted a run chart. Then they used SPC to identify some of the process problems and eliminated them. The thickness of the board became less variable almost immediately. Previously they had been continually adjusting the pressures and temperatures

Figure 11.1   Five stage improvement plan.

of the corrugating machine without any long term success, a strategy that in fact only made things worse.

In this chapter and in Chapters 12 and 13 we consider and discuss the following aspects of statistical process control:

- The benefits of reducing variation.
- The effect of tampering with a process. This is often done in practice but usually ends up increasing rather than reducing variation.
- The common cause highway, statistical control and stability.
- The distinction between special and common causes of variability. Understanding this distinction is the key to reducing variation.
- The construction and use of control charts to help identify common and special causes.
- The establishment of control charts and their use in monitoring processes.
- Specifications and capability. We need to examine whether a process is capable of meeting customer needs, as determined by process specifications.
- Strategies for reducing variation. Different strategies are needed to reduce common and special causes of variation.

11.2 Processes

The aim of statistical process control is to improve processes. What is a process? Processes cover everything that goes on in an organisation. In fact an organisation is made up of hundreds, maybe thousands, of interrelated processes. Anything that goes on is part of some process. Here are some examples:

> *Manufacturing:* Designing, assembling, packaging, maintaining equipment
> *Finance*: Payroll, accounts payable, accounts receivable, data processing, auditing
> *Employee relations:* Hiring new staff, training, salary reviews
> *Marketing:* Advertising, promotion, customer service, customer complaints, selling
> *Distribution:* Order taking, order filling, delivering, inventory management

A process is made up of three essential elements: inputs, processing system, and outputs. The flow of a process is shown in Figure 11.2. A processing system acts on the inputs by modifying them and adding value to produce useful outputs. The inputs may be materials (steel, electrical components, paint, paper,

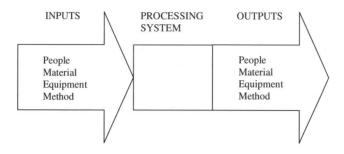

Figure 11.2   Elements of a process.

electrical power, etc.) but could also be people (ideas, work, etc.), equipment (machines, tools, etc.), and methods (systems, IT, etc.). Similarly, the outputs produced will fall into one or more of these classes.

Q1. What are the inputs and outputs of the beads experiment in Chapter 1?
Q2. How do you think processes are monitored and improved in most organisations? Think of an organisation with which you have been involved (school, university, vacation job, etc.).

Processes produce variable outputs. Variability arises from two sources: the processing system and the variability of materials used in the inputs. Figure 11.3 shows variability in inputs that, when added to the variability from the processing system, produces outputs that are often more variable.

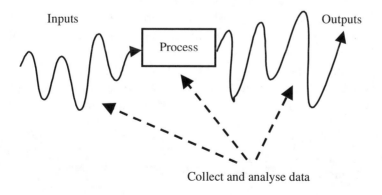

Figure 11.3   Variability in inputs, processing system, and outputs.

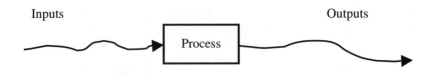

Figure 11.4   An improved process.

We improve a process by collecting and analysing data from the inputs and outputs and from studying the process itself. In particular, we concentrate on the outputs to see if we can identify causes of variability, and then we put in place solutions that address them. The problem solving techniques of Chapter 2 can be used to achieve this aim. By applying such methods, the process will be improved, resulting in less variability in the inputs, outputs, and the processing system, as depicted in Figure 11.4.

11.3 Benefits of Reducing Variation

Reducing variation is the key to improving quality, productivity, and profitability. Some of the benefits that can be gained from reducing variability include reduced costs, fewer errors, smaller inventories, shorter setup times, greater throughput, a more reliable and consistent service, and greater customer satisfaction.

🎬 *Dice Experiment*

To demonstrate the effects of reduced variability on throughput, we will consider a simple manufacturing system involving five processes: *material supplies, manufacturing, quality control, packing,* and *delivery.* At the end of the system a component is produced, which is then supplied to a customer. A schematic of the system is shown in Figure 11.5.

The customer requires 35 components every two weeks. The production line works 10 days every two weeks, so the average daily output needs to be 3.5 components per day. As in most production systems, the cost of production depends not only on labour and materials but also on levels of the work in progress (WIP).

We can simulate the manufacturing system by throwing a die at each step of the production process. The face value of the die determines the amount that can be worked on, provided enough material from the previous stage is

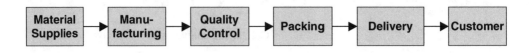

Figure 11.5   The manufacturing system.

Die	Face values	Mean	Standard deviation
1	1, 2, 3, 4, 5, 6	3.5	1.87
2	2, 3, 4, 5, 6, 7	4.5	1.87
3	3, 3, 3, 4, 4, 4	3.5	0.55

Figure 11.6   Experimental dice.

available. For instance, if a 6 is thrown but only three items are available from the previous stage, then only three items can be worked on.

Three different dice, which differ in the average values and variability of the six sides, will be used. These dice are listed in Figure 11.6. The first two dice have the same high variability but differ in average value. The third die has the same average value as the first die but much lower variability.

For simplicity, we assume that the supplier of the raw material is operating a *just-in-time* policy such that at the start of each day there are just enough materials, taking into account any unused materials from the previous day, to allow for the maximum possible production on that day. So, for example, if three units of raw material are left over from the previous day and if the maximum possible production level is six per day, then three further units of raw materials are made available.

At the beginning of each day, the first throw of the die gives the number of units that are *manufactured* on that day. Any excess of raw materials is held as WIP at the manufacturing stage and is available for use the following day. The second throw of the die determines the number of units that can be handled by *quality control*. That is, it gives the capacity of quality control for that day. If this capacity exceeds the number of units available from manufacturing, then all these units can be passed on to packing, with no WIP. On the other hand, if the capacity is less than the number of units received from manufacturing, then only this number can be passed on to packing, with the excess units being WIP at the quality control stage. The die is thrown two more times to represent capacity at the packing and delivery stages, where the same procedure is repeated so that finally the amount sent to the customer on each day is determined. This whole process is repeated for 50 days and with each of the three dice. Our contract is to supply the customer with an average of 3.5 components per day, that is, a total of 175 components over the 50 days.

Q3. What data should we collect from the simulations to compare the different patterns of production?

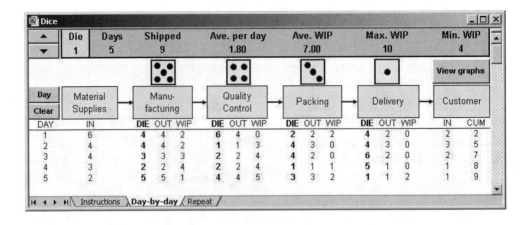

Figure 11.7   Dice experiment spreadsheet.

We suggest that you experience the dice experiment in a group situation. One option is to move material (such as pieces of paper or counters) from one process to another on the throw of a die. Another option is to simulate the experiment using Excel. Here we will use the spreadsheet *Dice.xls* to describe the experiment. This spreadsheet can be found on the CD-ROM.

The spreadsheet in Figure 11.7 shows the simulation of five days of production, on a *day-by-day* basis, using die 1. If we step through the results for day 1, we see that six pieces are available to work on in *manufacturing*. However, the 4 thrown by the die for manufacturing means they can only work on four of the six available pieces. These four are passed on to *quality control*, leaving the other two pieces as WIP. Although a 6 is now thrown in *quality control,* only four components are available to be worked on. The four are passed on to *packing,* leaving no WIP in *quality control.* Similarly, *packing* is only able to work on two of the four available, leaving two as WIP, and *delivery* is able to work on four, although only two pieces are available. At the end of the first day, therefore, the customer receives only two components. Day 2 proceeds in a similar way, but we must bear in mind that WIP at some stages is available from the previous day.

The first two rows of the spreadsheet record cumulative progress. Hence, after five days of production the customer has been shipped nine components, for an average of 1.80 per day. The total WIP on each of the five days is 4, 5, 7, 9, and 10, respectively, giving an average WIP of 7.0 pieces per day, with a daily maximum and minimum WIP of 10 and 4, respectively.

Each day's results are simulated using the *Day* button. Clicking on the button again gives the results for day 6. Repeatedly clicking this button will add further days. In this way the results for 50 days are obtained. The final cumulative figures will be given in the first two rows.

Clicking the *View graphs* button will show run charts of the WIP at manufacturing, quality control, packing, and delivery. Then clicking the *Throughput graph* button will display a run chart of the numbers of components shipped, with a horizontal line representing the average number.

Die	Components shipped	Average WIP per day
1		
2		
3		

Figure 11.8 Results from the simulation experiment.

The simulation can be repeated with different dice by first clicking the *Clear* button and then using the arrows in the top left-hand corner of the spreadsheet to choose a different die. For instance, clicking the up-arrow once will change *Die 1* to *Die 2*.

We suggest you run the experiment using *Dice.xls*. Record your results for 50 days of production for each of the three dice in Figure 11.8. (Alternatively, enter the data given in Figure 11.18 into Figure 11.8.)

The *Repeat* worksheet in *Dice.xls* can be used to simulate complete runs of the experiment for each of the three dice. Try running a few experiments to see what sort of variability you get in your results.

Q4. What conclusions do you draw from Figure 11.8?

Although your particular experience with the dice experiment may have its own idiosyncrasies, you should come to the following conclusions if you repeat the experiment many times:

- Increasing the average throughput of each process without reducing variation will tend to give the customer more than he or she needs. Thus, inventory increases dramatically, with all its associated costs. In particular, the finished goods inventory will be high, which often leads managers to continually change the level of production to adjust for oversupply. You could simulate this by switching between dies 1 and 2 while trying to maintain a balance between under- and oversupply. This could be viewed as equivalent to continually changing the number of staff working on the processes.
- Reducing the variation without increasing the average throughput of each process also increases throughput. However, this is not accompanied by increased inventory. In fact the inventory being held falls. Obviously if we could reduce variation so that each process produced exactly 3.5 units per day (perfect line balancing), there would be no inventory at all, and we would deliver exactly what the customer requires. Unfortunately, real production lines are much more complicated, and perfect line balancing like

this is difficult to achieve. Nevertheless, any reductions in variation will lead to increased throughput at a lower cost.

🏭 *Kanga Packaging*

A manufacturer of cardboard cartons was unable to meet the needs of one of its major customers during a particular period. In order to meet the demand the production manager arranged for the workers to work two hours overtime per day.

> **Q5.** *In light of the dice experiment, comment on this decision. What would be a better course of action?*

11.4 Common Cause Highway

As we pointed out in Section 11.2, most processes produce variable outputs. In many cases, this variability is inherent in the system and can only be reduced by making fundamental changes to the system. For example, the results from the red beads experiment in Chapter 1 showed considerable variation. This variability:

- Was due entirely to chance.
- Had nothing to do with the operators, so it was unfair to penalise them or give them bonuses for the results they achieved because the results were outside their control.
- Was a consequence of the *system*, namely, too many red beads in the raw material. The only way to get better results is to improve the raw material. Improving the system is the job of the manager.

A run chart of the results obtained from running the beads experiment 17 times, given in Figure 1.9, is reproduced in Figure 11.9. Remember, in each experiment five operators produce a total of 50 beads (the number in the paddle) for each of five days. This yields, therefore, the 425 individual results in Figure 11.9.

We can see that nearly all the data lie within the dotted lines on the run chart. If we repeated the experiment again, our data would almost certainly fall within these dotted lines. The beads experiment only generates what is known as *common cause variation*. We shall call the area within these dotted lines the *common cause highway*. We would be surprised if a result had fallen outside the common cause highway. Getting 0, 30, 40, or 50 red beads in the paddle would certainly surprise us. We would suspect that something unusual or *special* had occurred. Even 20 red beads would lead us to suspect that some special event had taken place. Note that a value has fallen outside the common cause highway

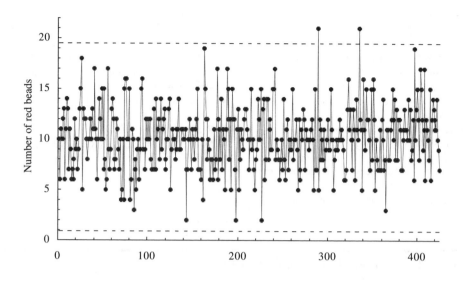

Figure 11.9   Results from the 17 beads experiments.

just twice in our 425 results, which is less than 0.5% of the time. Although possible, it is somewhat unlikely that these values were due to common cause variation. Thus it is suspected that they are due to the presence of some *special cause of variation*. Such values are always worth investigating further.

Similarly any obvious patterns in the data would surprise us. For example, if all the data in earlier experiments were below the overall average and all the data from later experiments were above the overall average, then we would suspect that something special had happened. Such a pattern would again be worth investigating to discover what had changed in the experiment. Maybe the number of red beads in the raw material had been increased. What constitutes an "obvious" pattern will be discussed in Chapter 12.

If all the data fall within the common cause highway and there is no "obvious" pattern in the data, we shall say that the data are *randomly distributed* on the common cause highway.

The key to understanding and then reducing the variation in any process is to be able to distinguish between:

- Data randomly distributed on the common cause highway, where variation can only be reduced by changes to the process, and
- Data that either fall outside this highway or have arisen from an "obvious" pattern, where variation can be reduced by a separate investigation of what are almost certainly special causes.

In Sections 1.2, 1.3, and 1.4 we looked at examples where variability in the data caused managers to make inappropriate decisions. Their difficulties were caused by not understanding the difference between common cause variation (data randomly distributed within the highway) and special cause variation. If the variation observed is common cause variation, as it probably was in every

Special causes of variation	Common causes of variation
• Localised in nature, such as a particular supply of raw material, a particular machine or a specific operator. • Not part of the overall system. • Not always present in the process, as they come from outside the usual process. • To be considered as abnormalities, or as unusual results, or as nonrandom patterns. • Things that can usually be put right by people working on the process, but sometimes needing management intervention. • Causes that typically contribute greatly to variation.	• Those that exist because of the processing system or the way the system is managed. • Present in the process all the time, although their impact varies. • Common to all machines, all operators, and to all parts of the process. • Random fluctuations in process outputs. • Events that individually have a small effect on variation but collectively can add up to quite a lot of variation.

Figure 11.10   Special and common causes.

case, then actions that treat certain results as in some way special are inappropriate. What is required is a more systemic remedy aimed at reducing the underlying level of variation in the results.

It is important to distinguish between common and special causes of variation. Techniques that enable us to do this comprise the subject area known as *statistical process control*. In Chapter 12 we shall see how to calculate the boundaries of the common cause highway to give us charts known as *control charts* that help us to distinguish between common and special cause variation. Once we know which type of variation we are dealing with, then we can employ the appropriate strategies discussed in Chapter 13 to reduce variation. Figure 11.10 gives an indication of the type of causes often associated with these two types of variation.

> *Q6. The results students obtain in different courses will naturally vary. Think of three or four reasons for this variation, and decide whether you consider them special causes or common causes.*

Epic Videos

Heather is the manager of the Chartwell branch of Epic Videos Ltd. All branch managers have recently been urged by head office to improve sales, and so Heather decides to collect data on the weekly sales of videos over several weeks. A run chart of the data she collected is given in Figure 11.11.

Most of Heather's data fall within the common cause highway, with no "obvious" patterns. Two data points fall outside the highway. There are two weeks where specific events caused sales to vary significantly. The low sales in week 12 were

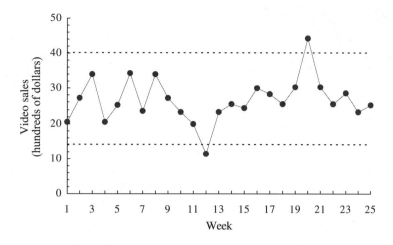

Figure 11.11   Video sales.

probably due to the occurrence of a public holiday that week, and the high sales in week 20 coincided with a TV sales drive. Note, however, the considerable variation within the highway. A number of common causes of variation contributed to this variation, such as weather, local sporting events, and TV programmes.

There appear to be two special causes:

- One of them is "good," and when we investigate it we find it is the result of a TV sales drive.
- The other is "bad" but is understandable on looking back through the records, for the public holiday probably caused it.

11.5 The 85/15 Rule

It has been established time and time again in companies throughout the world that at least 85% of all problems result from common causes of variation. That is, at least 85% of problems belong to the system and are thus the responsibility of management. Less than 15% are special causes.

Many believe the figure is considerably higher than 85%. W. Edwards Deming, a pioneer in the use of statistical process control and widely recognised for his prominent role in the recovery of Japanese industry after the Second World War, put the figure at 96%. The managing director of one prominent New Zealand manufacturing company believes the figure is 100%! That is, all the problems faced by the company in terms of excessive variation in their processes have ultimately been identified as being due to common causes of variation.

This is very different from the way most managers react. They are only too ready to blame people for problems. This is wrong. Most of the time (at least 85%), the problem is exactly like "too many red beads in the raw material" — a system problem.

> **Q7.** *Is it fair to appraise the performance of the operators involved with pro-*
> *ducing white beads? What do you think?*

11.6 Stability and Predictability

A process is said to be stable, or in *statistical control*, if no special causes are
present, that is, if no out of control points (all the data are on the common cause
highway) are present and the data are randomly distributed about the centre
line (with no "obvious" patterns).

There are a couple of advantages in having a stable process:

- The process is predictable in the long run, so the same pattern of variability
  will occur time and time again, and the customer knows what will be
  received. In contrast, with an unstable process, special causes dominate, so
  it becomes impossible to predict what will happen in the future.
- Nothing is gained by adjusting a stable process on the basis of its performance. If it is relatively easy to make adjustments continuously as information is collected, using, for instance, an automatic feedback system, it is
  tempting to tamper with the process. But, as we show in Section 11.7,
  adjusting a stable process will only make things worse. Once stable, the
  process cannot be improved without fundamental systemic changes.

11.7 Process Adjustment

As the dice experiment shows, considerable benefits can be achieved by reducing the variability in a process. Many organisations and managers try to achieve
these benefits by continually adjusting the process. They will institute changes
in the process when the results appear to be on the high side or low side, or
they will set up other procedures for dealing with change. They are attempting
to compensate for outputs that *appear* to be drifting away from some target
and, by so doing, to bring the process back on target — in other words, to
exercise what they believe is appropriate operational or managerial control.

To illustrate some of the effects of process adjustment, we will consider an
example based on a real-life manufacturing process.

🏭 *Eston Tubes*

Eston Tubes manufactures aluminium tubing for use in the construction of racing
bicycle frames. It produces 10-metre lengths of tubing using an extrusion process, and Bernard is employed to ensure that the extruder produces tubing of
the required 50 mm diameter. Tubing with a diameter of more than 2 mm from
target has to be scrapped. As each length is extruded a laser device automatically
determines the average diameter of the tubing, which is displayed on a screen.
Bernard has been told that if the screen display shows that tubing is off target,
he should try to compensate for this by adjusting a dial that increases or

decreases the size of the extrusion nozzle so that the next length of tubing will be acceptable.

> **Q8.** *This is an example of an operator using judgement to make regular adjustments to a production process. Is this to be encouraged, or can you see any drawbacks to what Bernard is doing?*
>
> **Q9.** *What do you think the long term effects of such practices might be?*

The consequences of actions such as those used by Bernard can be easily demonstrated by setting up an experiment to investigate what happens to the output from a common cause process when it is subjected to regular adjustment. A mechanical sampling device called a *quincunx* can be used to demonstrate these effects. A picture of a quincunx is given in Figure 11.12. The device has an adjustable funnel that feeds balls through a pinboard into a series of parallel slots. The lines of pins on the pinboard are offset so that a ball falling from the funnel will bounce off successive pins with an equal chance of going either left or right each time. Thus there is a greater chance

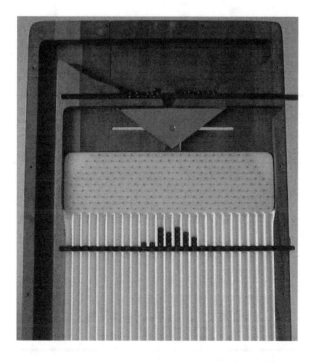

Figure 11.12   Quincunx.

that the balls will fall in the central slots than in the outer ones. If we drop a large number of balls, the slots fill up to form the shape of a distribution. Assuming we keep the funnel fixed in the central position, which is always targeting the middle slot, the distribution of balls will be close to a normal curve, with its mean near the centre slot. You can see this distribution beginning to form in Figure 11.12.

We can imagine that the middle slot of the quincunx represents the 50 mm target for the extrusion process. Slots to the left of the middle slot represent decreases of 1 mm and slots to the right increases of 1 mm. With the funnel positioned over the centre slot (50 mm), a ball is dropped from the funnel, through the pinboard and into one of the slots. This corresponds to the extrusion of a tube, with the resulting diameter given by the value of the slot containing the ball. If the funnel stays positioned over the 50 mm slot, this represents the situation where Bernard sets up the extruder to the best of his ability to produce 50 mm tubing, with no subsequent adjustment.

If you have a quincunx with an adjustable funnel, then it can be used to imitate Bernard's actions. Alternatively, you can simulate the experiment using the Excel spreadsheet *Quincunx.xls* given on the CD-ROM. Using either the quincunx or the spreadsheet, drop 25 balls without making any adjustment to the funnel (method 1 in the *Full Pinboard* worksheet), and plot your results in Figure 11.13. Alternatively, use the results in upcoming Figure 11.19 to answer Q10.

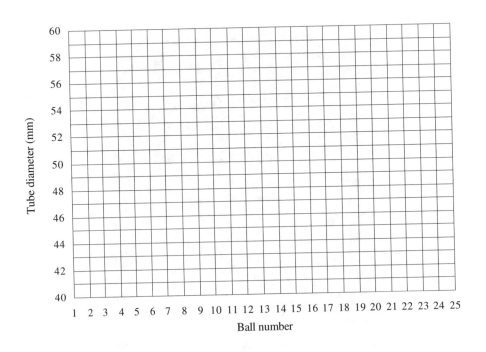

Figure 11.13   No adjustment.

Q10. *What proportion of the results will be scrapped, i.e., below 48 mm or above 52 mm?*

But this is not what Bernard is doing. He is trying to keep the process on target by continually adjusting the size of the nozzle. Suppose Bernard adjusts the process by moving the nozzle dial down 1 mm for every millimetre above 50 mm and up 1 mm for every millimetre below 50 mm. For example, if the screen shows 53 mm, he turns the dial down 3 mm for the next tube. Likewise, if the screen displays 48 mm, he turns the dial up 2 mm before the next tube is extruded.

This form of adjustment is called *adjusting for deviation*. It seems a natural way of trying to keep a process on target. After all, if a particular result is below the target, then we should surely make an upward adjustment. We can use the quincunx or spreadsheet to imitate the repeated adjustments made by Bernard by moving the funnel either to the left or right before each ball is dropped, depending on the result of the previous ball. Now drop 25 balls using this form of adjustment (method 2 in the *Full Pinboard* worksheet), and plot your results in Figure 11.14. Alternatively, use the results in upcoming Figure 11.20 to answer Q11 and Q12.

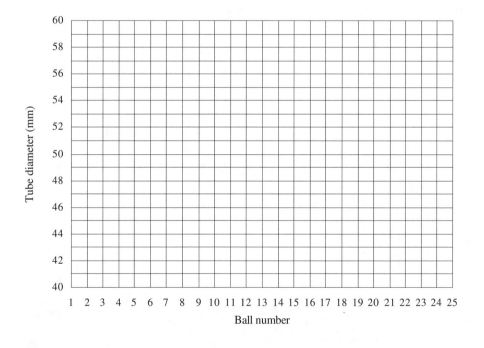

Figure 11.14   Adjusting for deviation.

> **Q11.** *In Figure 11.14, what proportion of tubes will be scrapped?*
> **Q12.** *Has Bernard improved the situation by continually adjusting the nozzle size? Why or why not?*

✎ *Tampering*

The type of process adjustment just described is an example of what is known as *tampering*. It can be done by anyone — line worker or manager — and even by machines. For a stable process, such actions are generally misguided and will usually make the situation worse, not better. If a process has only random variation, tampering with it will introduce even more variability into the process. Overreaction to random variation only increases the variation.

Tampering with a process occurs in many ways. Adjusting for deviation is a common practice, because it is natural to try to reset a process that seems to be deviating from some target. However, the form the adjustment takes can vary. For example, adjustment might only be made after two or three results have all fallen on one side of the target, or the amount of adjustment might try to compensate for a perceived bias in the output by setting the process "off-target" in the opposite direction to the run of results.

Another common form of tampering is to try to reduce variability by targeting the most recent results. This form of tampering, called *each one like the last*, is based on the belief that all variation can be controlled, and ultimately eliminated, by trying to produce complete consistency in a process. In the extreme, we might adjust after each output from the process, each time targeting the last value produced. More realistically, we might only shift to a new level when we get one or more values that deviate from the previous target by more than some amount. Although this form of adjustment is not common in the type of production process that Bernard is controlling, it does occur quite frequently in other business processes.

We can use the quincunx or spreadsheet (method 3 in the *Full Pinboard* worksheet) to simulate an *each one like the last* form of tampering. Before each ball is released, the funnel is positioned over the slot into which the previous ball fell.

> **Q13.** *What do you think will happen if we drop a large number of balls using this method of adjustment?*
> **Q14.** *Check whether you are right by using the quincunx or the spreadsheet (method 3) to simulate the effects of each one like the last. What do you see?*

In most industries tampering is a major source of increased costs. The remedies for dealing with excessive variation are to plot the data, stop tampering, and put in place a scientific approach to reduce variation.

Some Examples of Tampering

Consider the following examples of practices that are reasonably common in many companies. Each can be considered a form of tampering. What form of tampering is involved in each case? What do you think the long term effects of each action might be?

> **Q15.** *The sales department has a telephone budget of $10,000 per year, but the actual expenditure last year was $10,500. The manager decided to reduce next year's budget to $9500.*
>
> **Q16.** *The billing department has taken on a new employee, who is shown how to process bills by the most recently appointed member of staff. (After all it was not so long ago that she was shown the ropes.)*
>
> **Q17.** *In a car spraying process, batches of paint are mixed to spray around 30–40 cars the same colour. A sample of paint left over from the previous batch is used to match up the colour for the next batch.*

Improving the System

The only way to improve a system that has only common cause variation is to find and remove some of the common causes; strategies for doing this will be considered in Chapter 13. The quincunx can also be used to illustrate the consequences of removing common causes. So far we have used the full pinboard shown in Figure 11.15. Each row of pins introduces variability into the

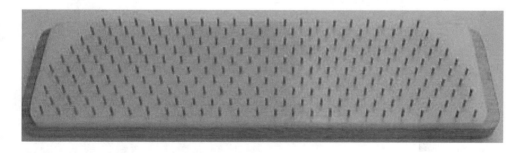

Figure 11.15   Full pinboard.

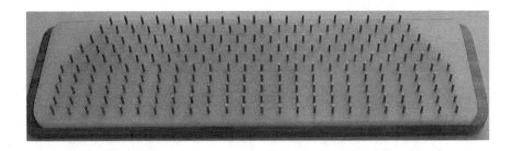

Figure 11.16   Half Pinboard.

results, since a ball can go to either side of the pin. Each row represents a common cause.

Suppose we were able to eliminate half the common causes. With the quincunx we can mimic this by using the pinboard shown in Figure 11.16. Note that the last five rows are in line, so when a ball reaches the sixth row it will continue straight through. If you use the spreadsheet *Quincunx.xls* choose the *Half Pinboard* worksheet.

The first strategy (method 1) involved leaving the funnel above the 50 mm slot at all times. We repeated this by dropping 500 balls using each of the two pinboards in turn. A box plot of the results is given in Figure 11.17.

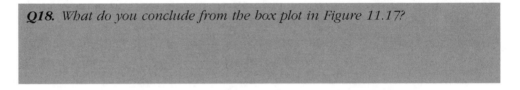

Q18. *What do you conclude from the box plot in Figure 11.17?*

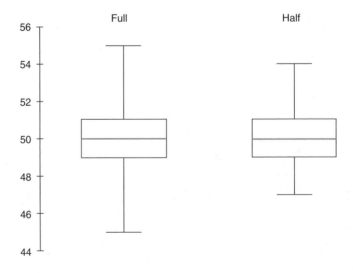

Figure 11.17   Results from both pinboards.

11.8 Chapter Summary

In this chapter we have covered the major ideas and philosophy behind statistical process control. Specifically:

- Improving business processes through reducing variation that results in:
  - Fewer defects and less rework
  - Improved productivity through better line balancing
- There are two types of variation:
  - common cause, where the data fall on the common cause highway with no "obvious" patterns. At least 85% of all problems arise from common causes of variation.
  - Special cause, where some data are outside the common cause highway or exhibit an "obvious" pattern within the highway. Less than 15% of problems arise from special causes of variation, which is contrary to how most managers react.
- Tampering is the nonscientific adjustment of processes and leads to increased variation.
- A process that is in statistical control, with no special causes present such that all the data are on the common cause highway with no "obvious" patterns, is a stable process.
- A stable process can only be improved by reducing or eliminating common causes of variation.

11.9 Data for the Dice Experiment

A set of sample results from 50 days of production for each of the three dice is given in Figure 11.18.

11.10 Data for the Quincunx Experiment

Figures 11.19 and 11.20 give results from dropping 200 balls in a quincunx. Those in Figure 11.19 were obtained when the funnel remained above the centre slot, representing the case of *no adjustment*. The results in Figure 11.20 are for the case when we were *adjusting for deviation*.

Die	Components shipped	Average WIP per day
1	145	15.76
2	197	20.64
3	164	4.94

Figure 11.18   Results from the dice simulation experiment.

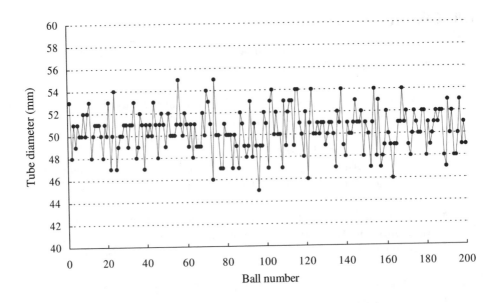

Figure 11.19   No adjustment.

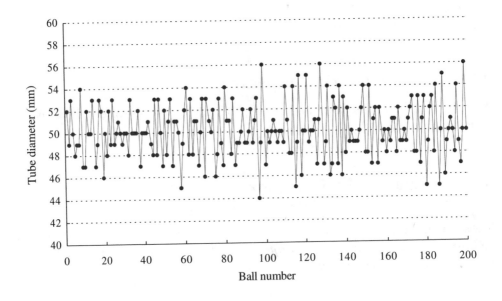

Figure 11.20   Adjusting for deviation.

11.11   Exercises

1. To control costs, project managers shall report on the costs of all projects that are 10% or more from the budget.
   a. What are your reactions to this policy?
   b. What do you think would be some of the effects of such a policy?

2. Last month it took two extra days to finish a project, so Mary arranged to provide these extra days for this month's project. Mary thought that if they needed two extra days, then they might as well have them. Do you believe her decision was sensible? Why or why not?

3. Applying and enrolling for a degree at a university is a process subject to variability. Some of the variation is due to special causes and others to common causes. Think back to your experience of this process or some other, similar experience. What things were good or bad, and on reflection do you think they were special causes or common causes?

4. A production line in the garment industry makes swimsuits. The line has several sewing processes that come one after the other. Outline the benefits that could arise from reducing the variation in process sewing times.

5. Wilson Hague is the personnel manager with Eldon Cycles Ltd. When he arrived at the company, he started a training programme, part of which is a two-day programme where new employees are shown how to operate several machine processes by other process workers. Discuss the merits of such a scheme.

6. Outline the major differences between special- and common cause variation. Why do you think common cause variation is more prevalent than special-cause variation in most business processes?

7. What is a stable process? Outline the advantages that arise from stable processes.

Chapter 12

Control Charts

12.1 Introduction

A control chart is a run chart on which the borders and centre line of the common cause highway have been drawn. In this chapter we consider how to construct control charts for different situations and types of data. We also discuss how to interpret the charts and how to use them to monitor a process. There are two main types of control charts:

- Variables control charts
  - The *individuals* chart: A chart used for interval data, when it is feasible to get only one measurement in any time period. For example, in the debt recovery example in Section 1.3, data on the percentage of unrecovered debt was available only once a month.
  - The *R* and *X-bar* charts: Two control charts used for interval data, when a number of measurements can be obtained in a relatively short period of time. They are commonly used on data from the production floor, for example, the width of corrugated cardboard from a continuous process in the packaging industry, and the weight of cans of milk powder from a filling process in the dairy industry.
- Attribute control charts
  - The *p-chart*: A chart used for proportions, where samples of a fixed size are taken regularly and the proportionate occurrence of some attribute is determined, for example, the proportion of leaky cartons used for orange drinks, or the proportion of faulty microchips produced by a computer component manufacturer.
  - The *c-chart*: A chart used for count data, where we observe regularly the number of times that an event or feature occurs, for example, the number of accidents or machine breakdowns, or the number of burrs in reels of woven cloth.

Routines for drawing each of these control charts are provided in the *STiBstat* add-in. First, however, we look at some of the ideas and common themes behind the construction of control charts and define what constitutes "obvious" patterns that can occur on the common cause highway as a result of special causes of variation.

12.2 Three Sigma Limits

The common cause highway is constructed from process data, using estimates of the process mean and standard deviation. Specifically, it comprises a centre line, which represents the *location* of the data, and two control limits that reflect the typical *spread* of the data. The centre line is given by the arithmetic mean of the data, and the control limits are based on *sigma*, which is an estimate of the process standard deviation. If we add three times sigma to the mean, we get the upper boundary of the highway, known as the upper control limit (UCL). Subtracting three times sigma from the mean gives the lower boundary of the highway, or lower control limit (LCL). Thus, the common cause highway lies within three sigma of the mean, and we refer to the limits as *three sigma* limits. In Sections 12.5, 12.6, and 12.7 we shall be concerned with how we calculate these control limits and, in particular, how we obtain sigma. For now we shall discuss some of their features.

The first question to ask is why three sigma? Why have we multiplied by 3? With three sigma limits, if a result falls outside these limits, then it is almost certainly a special cause. That is, if a result falls outside the common cause highway, we have objective evidence that a special cause is present, and we can then look into this. Even with a process with no data outside the control limits, it is still possible that certain results are influenced by special causes; it is just that the results are not sufficiently different for this to be clear. Likewise, the chances are small that data will fall outside the control limits even if no special causes are present. For instance, with the beads experiment it is possible to get 0 or 25 or even 50 red beads in the paddle. But the chances are very small. In fact, with a common cause process we would expect about 2 results out of 1000 to fall outside the three sigma limits purely by chance.

But why three sigma limits? Why not two sigma limits or four sigma limits? We shall see in Chapter 13 that we adopt very different strategies for dealing with common and special causes of variation. common cause variation indicates a system problem. Special cause variation is more indicative of a one-off event that requires immediate and special attention. It is important that we do not confuse these two types of causes, that is, treating a result as if it were a special cause when it is not, and vice versa. We can identify two mistakes that such confusion causes:

- *Interfering too often in the process.* Thinking that the problem is a special cause when in fact it belongs to the system. If we use two sigma limits, we shall find that we have many results outside the limits (about 1 in 20) when we have a common cause process.
- *Missing important events.* Saying that a result belongs to the system when in fact it is a special cause. With four sigma limits, we shall miss the opportunity to attack serious special causes.

The use of three sigma limits is thus a compromise, which has proved itself time and time again in practice over the past 50 years. In other words, *it works!*

12.3 Special Cause Patterns

In Section 11.4 we said that for a process without any special causes, the results should vary randomly about the centre of the common cause highway. That is, virtually all the data fall within the common cause highway and with no "obvious" patterns in the data. However, specific patterns on the control chart may indicate a lack of randomness or a lack of stability and suggest the need to look for special causes. Lloyd Nelson ("The Shewart Control Chart: Tests for Special Causes," *Journal of Quality Technology*, 16, 237–239 1984) has given a number of tests based on patterns of points that would be unexpected with a common cause process. Each signals that something unusual is happening. We consider some of these patterns in Figure 12.1.

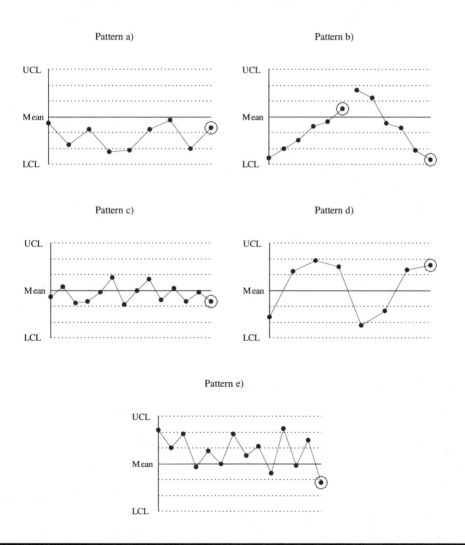

Figure 12.1 Patterns in control charts.

The dotted lines are at *one, two,* and *three sigma* above and below the centre line of the common cause highway. Specifically:

■ Figure 12.1a shows a long run of points below the mean, indicating that the mean may have shifted down. A sequence of nine consecutive points below the mean is a special cause signal. Obviously, the same situation applies when all the points are above the mean.

■ The situation in Figure 12.1b, with runs either upward or downward, may indicate a trend in the results. A sequence of six consecutive points steadily increasing or decreasing provides the signal. Such trends may be explained, for instance, by the presence of a learning effect, the wearing down of a tool, a move to improved materials, or improved operator training.

■ In Figure 12.1c the results are more tightly clustered about the centre line than would be expected with a common cause process. This could be due to incorrectly drawn control limits, perhaps from not recalculating the limits after some change in procedure had reduced the variability in the process. However, it may also be a result of inspection where results outside the specification limits are not plotted, thus giving a misleading picture of what the process is capable of producing. A sequence of 14 consecutive points within the one-sigma limits is needed for a signal.

■ In contrast, no points are close to the mean in Figure 12.1d. The cause here is probably failure to stratify. The results above the mean, for instance, may come from one batch of raw material, the results below the mean from a different batch. There are two separate processes involved. The signal here is a sequence of eight consecutive points all falling outside the one-sigma limits.

■ The systematic sawtooth effect in Figure 12.1e may be due to a failure to stratify, for instance, where the process has been sampled from two different processes alternately. Alternatively, it could be due to overadjustment of the process. A sequence of 14 consecutive points alternating up and down provides a signal.

Note that people are very good at finding patterns in data where none really exist. It is important not to get carried away. The foregoing tests give a set of *objective* criteria to signal the presence of what is likely to be a real underlying pattern or special cause.

12.4 Variables Control Charts

Variables control charts are used when interval data are collected on the process. If the only data available on the process are a single measurement taken at some regular time period, then an *individuals chart* is used. Often all we can get is one measurement each day, week, or month. Production figures, sales, and financial data are some examples. The debt recovery data in Section 1.3 and the budget deviations data in Section 1.5 are of this form. The individuals chart will give us information about how the process changes over time. If special causes are present, then the chart should alert us to this. Individuals charts will be considered in Section 12.5.

In other situations it will be possible to collect a small sample (or *subgroup*) of measurements from the process over a relatively short period of time, say, a few minutes, and to repeat the sampling at regular intervals, such as every hour. For instance, the shrinkage of a plastic product from an extrusion process, the weight of blocks of cheese in a dairy factory, and the thickness of plywood sheets could be sampled and measured in this way. Now we can see not only whether the mean of each subgroup changes from, say, hour to hour but also whether the subgroups are consistent from hour to hour. That is, whether the variation *within* our subgroups is due to common causes or whether special causes are present. Two charts are, therefore, required to study the variation within and between the subgroups, an *R chart* and an *X-bar chart*. These charts will be discussed in Section 12.6.

If a subgroup of items can be collected, then more information about the process can be obtained. However, care has to be taken in deciding what constitutes a subgroup. For instance, suppose production data are available for each of the five days of the workweek. If these data are thought of as a subgroup, then the use of an *R* chart and an *X*-bar chart would allow us to examine the variability in production both within weeks and from week to week. However, the basic idea in forming a subgroup is that, if a special cause is present, it should affect *all* of the items within the subgroup in the same way. That is, variation within a subgroup should be due to common causes alone. It is probably unrealistic to assume that daily production figures within any particular week are relatively stable and will only be subject to common cause variation. Hence, it will be more appropriate to use an individuals chart for such data. This issue is discussed further in Section 12.9.

12.5 The Individuals Chart

In order to establish an individuals control chart, we need to collect at least 20 measurements from the process in which we are interested. The *three sigma* limits are calculated from the standard deviation of the data to give the control limits as

$$\text{LCL} = \text{Mean} - (3 \times \text{Standard deviation})$$

and

$$\text{UCL} = \text{Mean} + (3 \times \text{Standard deviation})$$

The control chart is obtained by drawing these limits and the centre line (the mean) on a run chart of the individual values.

Care has to be taken when calculating the standard deviation. If special causes have occurred during the data-collection period, the size of the standard deviation can be affected and, hence, the width of the control limits. If any special causes are present, the standard deviation could be larger than it would be otherwise. In this case, it would probably be sensible to calculate the standard deviation with the special causes omitted. Other methods of calculating the variability based on a *moving range* have been suggested; see, for example, Joiner's *Fourth Generation Management* (p. 234). These methods are less sensitive to *outliers*, i.e., to points that are likely to be outside the common cause highway.

📈 *Debt Recovery*

In the debt recovery problem discussed in Section 1.3, a run chart of the percentage of debts collected within the due month was given in Figure 1.2. It was shown in Sections 4.3 and 4.4 that the mean and standard deviation of these data were 76.53% and 4.26%, respectively. Hence, the control limits are

$$\text{LCL} = 76.53 - (3 \times 4.26) = 63.75 \quad \text{and} \quad \text{UCL} = 76.53 + (3 \times 4.26) = 89.31$$

The resulting control chart is given in Figure 12.2.

Q1. What do you conclude from this chart?

It can be seen that January figures are consistently lower than for other months. A reason for this has been established, that (in the southern hemisphere) January is the month when most businesses close down for summer vacation.

Q2. Recalculate the control limits and centre line with the January figures omitted, and plot them in Figure 12.2. (The full set of data is given in Figure 4.1.)
Q3. Are your conclusions any different now?

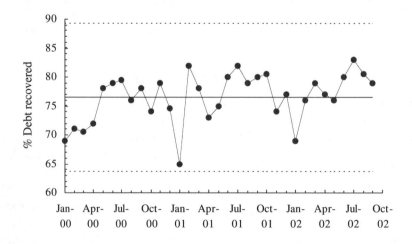

Figure 12.2   Percentage of debts collected within due month.

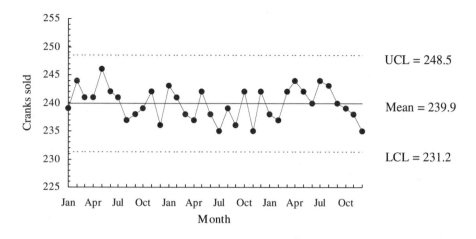

Figure 12.3   Control chart for cycle crank sales.

📈 *Eldon Cycle Cranks*

Jim Boardman, the sales manager of Eldon Cycles, has doubts about whether the company's new 75 mm cycle cranks are continuing to sell since their major competitor brought its own product onto the market last July. Consequently, Jim has collected data on the number of cranks sold per month for the last three years. He has calculated the control limits and constructed the control chart given in Figure 12.3.

> **Q4.** *What should Jim conclude from his analysis?*

12.6 *R* and *X*-Bar Charts

In order to discuss some of the issues involved in the construction of *R* and *X*-bar charts, consider the following example from the plastics industry.

📈 *Nuva Plastics*

Ellen Vagner, a line manager with Nuva Plastics, is responsible for the process that produces moulded laundry baskets. Each morning the company starts the process up at 8:00 a.m., and the die that moulds the baskets is heated to 400°C for half an hour before the first basket is moulded at 8:30 a.m. Production continues throughout the day until the process is closed down at 6:00 p.m. Recently Nuva has received some complaints about the strength of the baskets. It is well known in the industry that strength is mainly related to the density of plastic used in the die, so

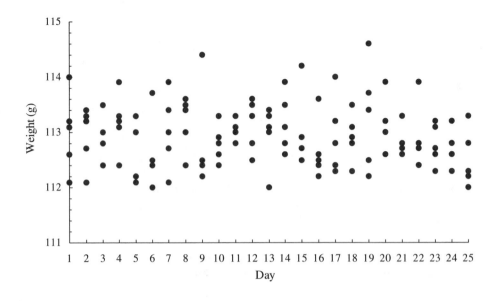

Figure 12.4   Weight of laundry baskets.

the company decided to monitor the weight (in grams) of the baskets after they have been moulded. It collected five samples, spread evenly over a day, and continued collecting data for a period of 25 days. The data are plotted in Figure 12.4. If fewer than five points appear on any day (for instance, on day 3), this is because two (or more) values coincide at the same point on the graph.

If we look at the daily averages of the data in Figure 12.4 we can see, without doing any calculations, that some variation exists *between* days. For example, the average for day 4 is higher than that for days 2 and 3, and the average for day 25 is lower than that for the preceding four days. Are these daily averages within the common cause highway? That is, is the variation in the daily averages due simply to chance (common cause variation), or is the average greater or smaller on some days than what might be expected (special cause variation)? Do any patterns exist in the way the daily averages vary? The *X-bar chart* for the daily averages will help us answer such questions.

What can we say about the variation *within* a day? We can see, for instance, a greater range of weights in day 19 than in day 11, where there is very little variation in the five samples. We can construct the *R chart* of the daily variation, as measured by the daily range. This will give us useful information about whether the weights within our samples are consistent from day to day.

✥ *Constructing R and X-Bar Charts*

The variability in the data *within* each subgroup will provide the basis for an estimate of the variability in the process. This estimate will be less sensitive to special causes than an estimate that does not differentiate between subgroups. In previous chapters, we have mainly used the standard deviation to measure variability. However, where small samples of data are being taken regularly, it is much simpler and just as effective to use the range as a measure of variability.

The steps involved in the establishment of the R and X-bar charts are as follows:

1. Collect 20 or more subgroups of data. Each subgroup should have the same number of data values (usually three to six but most commonly five). Any unusual events should be recorded.
2. For each subgroup, calculate the mean and range.
3. On separate run charts, plot the ranges (R chart) and means (X-bar chart).
4. Calculate the mean of the subgroup ranges, denoted by $\bar{R}$. The *three sigma* limits for the R chart are given by

$$\text{LCL} = D_3 \bar{R} \quad \text{and} \quad \text{UCL} = D_4 \bar{R}$$

   where the constants D_3 and D_4 are as given in Appendix A.6.
5. Look for points outside the control limits on the R chart. If they are identifiable as special causes, they can be removed from the data. The control limits should then be recalculated.
6. Calculate the mean of the subgroup means. Denote this by $\bar{\bar{x}}$. The *three sigma* limits for the X-bar chart are given by

$$\text{LCL} = \bar{\bar{x}} - A_2 \bar{R} \quad \text{and} \quad \text{UCL} = \bar{\bar{x}} + A_2 \bar{R}$$

   where A_2 is also as given in Appendix A.6.
7. Look for points outside the control limits on the X-bar chart. Try to find special causes for these points, in which case they should be removed and the control limits for both charts recalculated.

Note that the constants in Appendix A.6 depend on the size of the subgroup chosen. For example, if we are using subgroups of size 5 then

$$A_2 = 0.577 \quad D_3 = 0, \quad D_4 = 2.114$$

These limits are three sigma limits, although this is not apparent from the formulae. Statistical theory tells us that for subgroup sizes less than 10 the range provides a good estimate of the variation in the data. It is not the same as the standard deviation, but it is related to it. This relationship depends on the subgroup size, and this is reflected in the constants given in Appendix A.6.

Nuva Plastics

The plot of the data collected by Ellen Vagner on the weights of moulded laundry baskets is shown in Figure 12.4. The full set of data for 25 days is in cells B2:F26 of the Excel spreadsheet in Figure 12.5. The subgroup means and ranges are calculated in columns G and H. The formula in cell G2 is entered as

$$= \text{AVERAGE(B2:F2)}$$

	A	B	C	D	E	F	G	H
Nuva Plastics								_ □ ×
1	Subgroup	Item1	Item2	Item3	Item4	Item5	Mean	Range
2	1	114.0	112.6	113.2	113.1	112.1	113.0	1.9
3	2	113.2	113.3	112.7	113.4	112.1	112.9	1.3
4	3	113.5	112.8	113.0	112.8	112.4	112.9	1.1
5	4	113.9	112.4	113.3	113.1	113.2	113.2	1.5
6	5	113.0	113.0	112.1	112.2	113.3	112.7	1.2
7	6	113.7	112.0	112.5	112.4	112.4	112.6	1.7
8	7	113.9	112.1	112.7	113.4	113.0	113.0	1.8
9	8	113.4	113.6	113.0	112.4	113.5	113.2	1.2
10	9	114.4	112.4	112.2	112.4	112.5	112.8	2.2
11	10	113.3	112.4	112.6	112.9	112.8	112.8	0.9
12	11	113.3	112.8	113.0	113.0	113.1	113.0	0.5
13	12	113.6	112.5	113.3	113.5	112.8	113.1	1.1
14	13	113.4	113.3	112.0	113.0	113.1	113.0	1.4
15	14	113.9	113.1	113.5	112.6	112.8	113.2	1.3
16	15	114.2	112.7	112.9	112.9	112.5	113.0	1.7
17	16	113.6	112.6	112.4	112.5	112.2	112.7	1.4
18	17	114.0	113.2	112.4	112.3	112.8	112.9	1.7
19	18	113.1	112.9	113.5	112.3	112.8	112.9	1.2
20	19	114.6	113.7	113.4	112.2	112.5	113.3	2.4
21	20	113.9	113.0	113.0	113.2	112.6	113.1	1.3
22	21	113.3	112.7	112.6	112.8	112.7	112.8	0.7
23	22	113.9	112.4	112.7	112.4	112.8	112.8	1.5
24	23	113.2	112.3	112.6	113.1	112.7	112.8	0.9
25	24	113.2	112.8	112.8	112.3	112.6	112.7	0.9
26	25	113.3	112.8	112.0	112.3	112.2	112.5	1.3
27							112.92	1.36

| ◄ ◄ ► ►| \Sheet1 / Sheet2 / Sheet3 / | ◄ | ► |

Figure 12.5   Nuva Plastics data.

and the range in cell H2 is calculated from the difference between the maximum and minimum:

$$= MAX(B2:F2) - MIN(B2:F2)$$

These formulae are then copied to cells G3:H26. The overall mean and the mean range are calculated in cells G27 and H27 using the *AVERAGE* of G2:G26 and H2:H26, respectively.

> **Q5.** *Calculate the control limits for the R and X-bar charts, and plot them on the charts in Figures 12.6 and 12.7. Add the centre lines as well.*
>
> **Q6.** *Which control chart should you examine first, and why?*
>
> **Q7.** *What conclusions do you draw from your control charts?*

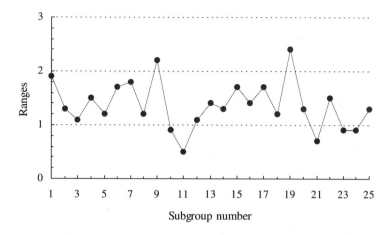

Figure 12.6   *R* chart for Nuva Plastics.

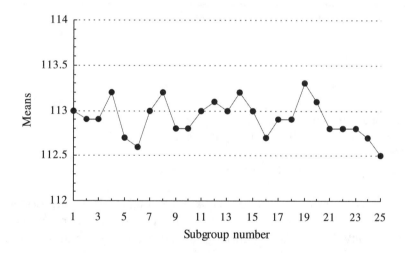

Figure 12.7   X-bar chart for Nuva Plastics.

✍ *Interpreting the* R *and* X-bar *Charts*

The *R chart* monitors the spread or variability of a process and is used to detect changes in the variability of the process. If the variation of the process changes, we would expect the chart to signal a special cause. Although the charts can signal increases or decreases in variability, it is usually points above the UCL that are of most concern. A point above the UCL on the *R* chart signals an increase in variability and may be due to such things as:

- ■ Tool wear
- ■ Poorly trained or fatigued operators

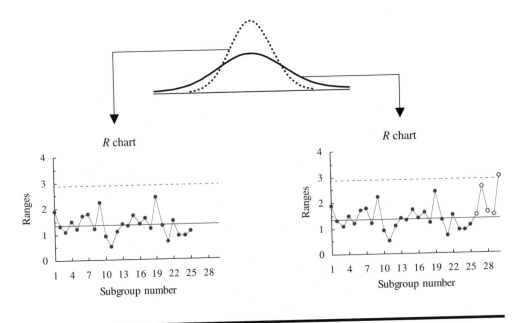

Figure 12.8   Detecting changes in variability.

- Start-up effects
- Increased rate of production
- Damaged tooling or fixture
- Poorer quality raw material or an end of lot
- Mixtures of raw material
- Automatic controller failure (overadjustment)

For example, suppose a process is sampled initially from the dotted distribution in Figure 12.8. After the 25th data point, the process changes to one with greater variability. We are now sampling from the solid distribution. The change is picked up in the *R*-bar chart after a further five points.

The *X*-bar chart detects shifts in the general level of the process. A signal from the chart may be due to such things as:

- An adjustment of tooling or fixtures
- A change in raw material, operator, or procedure
- A change in operating conditions
- A failure of the measurement system

Again suppose the process is sampled initially from the dotted distribution in Figure 12.9. After the 25th data point the process mean shifts upward, and we are now sampling from the solid distribution. The change is soon picked up in the *X*-bar chart. If the *R* and *X*-bar charts are both out of control, first examine the *R* chart. The *X*-bar chart can be affected by instabilities in the *R* chart.

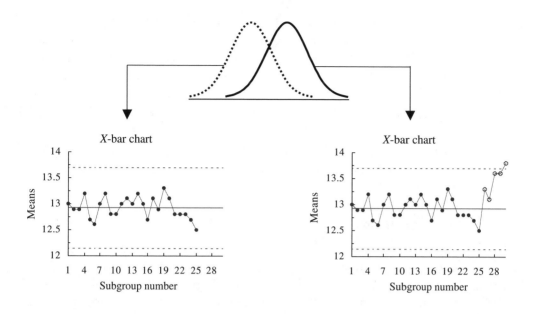

Figure 12.9 Detecting shifts in the mean.

12.7 Attribute Charts

We now consider two other types of control chart, known as *attribute control charts,* for data that are either proportions or counts. The charts are:

- The *p*-chart for the proportionate occurrence of some attribute, for example:
  - the proportion of ballpoint pens in a box of 20 that do not write
  - the proportion of plastic milk containers in a carton of 100 that leak
- The *c*-chart for the number of times that an event or feature occurs, for example:
  - the number of paintwork blemishes on a finished car
  - the number of typing errors per page

In calculating the three sigma limits for these two control charts, we shall use results for proportion data given in Section 4.12 and for counts data in Section 4.13.

✎ *The p-Chart*

The steps to follow in the construction of the *p*-chart for some attribute, such as a defective item, are as follows:

1. Collect at least 20 subgroups of *n* items each, and calculate the proportion (*p*) of each subgroup that possesses the relevant attribute. The subgroup size here should be larger than for *X*-bar and *R* charts and in general sufficient to give rise to at least a few items possessing the attribute.

2. Calculate the average proportion ($\bar{p}$) possessing the attribute.
3. Plot the proportion in each subgroup and the centre line ($\bar{p}$) on a run chart.
4. Calculate the *three sigma* control limits and add them to the chart. The limits are given by

$$\text{LCL} = \bar{p} - 3\sqrt{\frac{\bar{p}(1-\bar{p})}{n}} \qquad \text{and} \qquad \text{UCL} = \bar{p} + 3\sqrt{\frac{\bar{p}(1-\bar{p})}{n}}$$

5. Check for any points outside the control limits. If a special cause can be found, eliminate these points and recalculate the centre line and the control limits.

Dotoya Cars Ltd.

Dotoya Cars Ltd. assembles cars for sale in Australia. Recently it has been having problems with the production of rear seats, with some of the laminated plastic coming away from the seams. The company examined this problem over a period of 30 weeks. Each week it took a random sample of 40 seats and recorded the number with defective seams. The results are given in Figure 12.10 and plotted in Figure 12.11.

Q8. *Calculate the control limits and draw them on the graph in Figure 12.11.*
Q9. *What can you say about this process?*

Week	Number of defective seats	p	Week	Number of defective seats	p	Week	Number of defective seats	p
1	3	0.075	11	9	0.225	21	4	0.100
2	6	0.150	12	4	0.100	22	7	0.175
3	3	0.075	13	6	0.150	23	3	0.075
4	4	0.100	14	3	0.075	24	6	0.150
5	2	0.050	15	2	0.050	25	8	0.200
6	4	0.100	16	3	0.075	26	5	0.125
7	3	0.075	17	6	0.150	27	2	0.050
8	7	0.175	18	3	0.075	28	4	0.100
9	4	0.100	19	6	0.150	29	7	0.175
10	4	0.100	20	6	0.150	30	4	0.100
Totals	40	1.000		48	1.200		50	1.250

Figure 12.10   Numbers of defective seats.

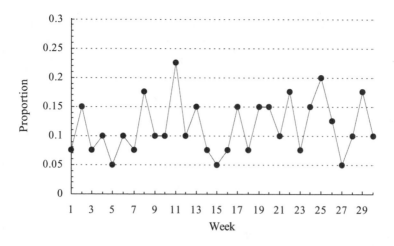

Figure 12.11   Proportion of defective seats.

✍ The c-Chart

The steps to follow in the construction of the c-chart for some event or feature, such as a defect, are as follows:

1. Obtain at least 20 similar inspection units, and count the number (c) of times that the event or feature occurs. Note that the inspection unit has to be large enough to produce observable effects.
2. Calculate the average number of occurrences per inspection unit, $\bar{c}$.
3. Plot the number of occurrences per inspection unit and the centre line ($\bar{c}$) on a run chart.
4. Calculate the three sigma limits, and add them to the chart. The limits are given by

$$\text{LCL} = \bar{c} - 3\sqrt{\bar{c}} \quad \text{and} \quad \text{UCL} = \bar{c} + 3\sqrt{\bar{c}}$$

5. Check for any points outside the control limits. If a special cause can be found, eliminate these points and recalculate the centre line and the control limits.

🏭 Dotoya Cars Ltd. Revisited

After consulting a statistician, Dotoya Cars Ltd. decided to look at its defective car seat problem again. The statistician advised the company it would be better to set up a c-chart, where the number of seams that are defective on a seat are counted, rather than classifying seats as defective or nondefective.

> *Q10. Why has the statistician given this advice? What are the merits of this approach, compared to the p-chart approach?*

Seat	Number of defective seams	Seat	Number of defective seams	Seat	Number of defective seams
1	7	11	8	21	6
2	5	12	7	22	6
3	0	13	4	23	4
4	8	14	6	24	7
5	3	15	2	25	7
6	6	16	7	26	3
7	5	17	5	27	3
8	4	18	5	28	3
9	5	19	3	29	3
10	5	20	5	30	3

Figure 12.12   Numbers of defective seams.

Data collected from 30 seats are given in Figure 12.12.

Q11. Construct the control limits for this example, and plot them on the chart in Figure 12.13.

Q12. What can you say about the stability of the process? Is the process in statistical control?

12.8 Monitoring the Process

Once appropriate control charts have been established, they can be used for monitoring by adding further points to the charts for each subsequent period as additional data are collected. This task is usually the responsibility of the process operators and supervisors, so they can get immediate feedback.

Prompt reaction to out of control points is essential for successful process improvement specifically:

■ It is much easier to find special causes on the spot, if at all possible, than to reconstruct the circumstances later.

■ Charts are very conservative. When a signal occurs it is very likely that a special cause is acting, presenting an opportunity to better understand and improve the process.

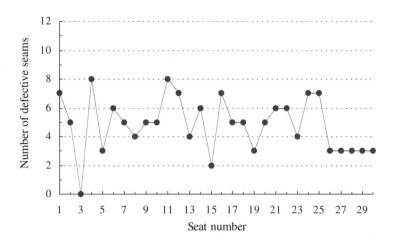

Figure 12.13   Number of defective seams on car seats.

- Lack of reaction causes all involved to doubt the usefulness of the control charts.
- It is better to have no chart at all than to ignore the signals from a chart in place.

Resources must be made available to search for special causes once a signal occurs. The project team should review the charting process periodically. If no signals are occurring, it is wise to confirm that the data are being collected according to plan. Consideration can be given to changing the nature of the subgroups, the frequency of sampling, and so on. If the variation in the process has been reduced by removing special causes, the control limits should be revised on the basis of recent data.

Dotoya Cars Ltd. Revisited

In the previous section, data collected on 30 seats were used to establish the c-chart given in Figure 12.13. The chart is now used to monitor the process. Data collected from the next seven seats had 3, 3, 2, 3 1, 0, and 3 defective seams.

Q13. *Plot these data on the control chart in Figure 12.13. What conclusions do you now draw?*

Q14. *What would you suggest Dotoya does next?*

12.9 Sensitivity of Control Charts

A forest products company constructed a control chart for one of its processes, but when it used the control chart for monitoring the process it discovered that the chart signalled special causes every hour. The company could not afford to stop the process because of lost production. Neither could it provide resources and time for operators to investigate the process every hour. It concluded that the control charts caused more problems than they solved, so they were discontinued.

What the company failed to understand was the sensitivity of the chart. The chart was signalling too often for the resources that were available. Charts can be made less or more sensitive by changing the frequency of sampling and/or the size of the subgroup sample. Generally, the more frequently a process is sampled, the larger the size of the subgroup, or both, the more sensitive the chart.

Consider again the Nuva Plastics example in Section 12.6. The line manager, Ellen Vagner, took samples of five baskets at five different times of the day.

> **Q15.** *In establishing the chart, do you think the subgroup size and frequency of sampling are appropriate? What can you say about the sensitivity of this chart?*
> **Q16.** *What are the possible dangers of sampling five times a day?*
> **Q17.** *How is the sensitivity of the chart affected by increasing the frequency of sampling to five per hour?*

12.10 Implementing Statistical Process Control

A project team should be responsible for *establishing* the chart. A team approach is useful especially in the identification of special causes. Once the chart is in place, responsibility can then be turned over to the operators. To ensure their cooperation, it is important to have operators involved in the project team. Basic training is also necessary so that everyone understands the purpose and use of control charts.

Placing a chart at the end of a complex process is ineffective because it is usually only possible to find special causes at the point where they occur. Charts should be placed at critical points within a complex process to make the task easier; that is, the process is best broken down into simple components.

It is much more informative to collect interval data than attribute data. Much less data is required to spot out of control conditions if data are measured on some appropriate interval scale rather than being simply recorded on a pass/fail or good/bad basis.

Data collected from a process can have variation that arises from two sources: the process itself and the measurement system. If the variation that arises from measurement is so large that it hides the variation in the process, then it will be pointless trying to establish a control chart until the measurement system is reliable. Thus, a preliminary study of the measurement system is advisable.

Quite often control charts fail in companies, not because of the technicalities, but because of a poor understanding of these and other management issues.

Assessing whether a process is capable of meeting customer needs and strategies for improvement are covered in Chapter 13.

12.11   Case Study: Mike's Orange Company Ltd.

Mike's Orange Company Ltd. produces fresh orange juice that is sold in cardboard cartons. Recently, retail customers have been complaining about the number of cartons that have leaks when they are put out on display. Gillian Munday, the line supervisor for the filling line, is given the task of investigating and eliminating this problem. She collects a random sample of 500 cartons from the process and examines them for leaks. She finds that 8% show signs of leakage, of which over 30% are leaking from the bottom left-hand front corner of the carton. On investigation she discovers that the machine that fills the cartons has a slight protrusion corresponding to this position, and she calls in the process engineer to fix the problem. Realising that she has found a possible reason for only 30% of the leaky cartons, she decides to monitor the process by setting up a control chart.

Q18. Which type of control chart do you think is most appropriate for monitoring the filling process?

Q19. What is the smallest subgroup size you would recommend for setting up the control chart?

Gillian decides to collect a sample of 50 cartons each morning and afternoon for the next 15 days to set up a control chart. A *p*-chart of her data is given in Figure 12.14. The mean percentage of leaky cartons over the 15-day period is 4.1%.

Q20. Verify the control limits on Figure 12.14. What would you do about the out of control point on the chart?

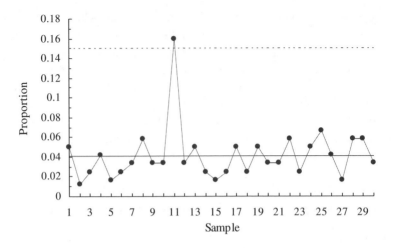

Figure 12.14   Leaky orange cartons.

The control chart is used to monitor the process. One morning the filler process signals a special cause, with far more leaky cartons than normal. Gillian examines the cartons and discovers that most of the leaks are from the seal on the top of the carton. Further investigation reveals that the new seals used that morning are smaller than usual and had been sourced from a different supplier. As a result, they decide to revert to their normal supplier, and a new batch of seals is ordered immediately.

> **Q21.** *What should Gillian do next to continue monitoring the process in her drive to reduce the number of leaky cartons?*

When the replacement seals arrive, a new chart is established showing a mean of 3.3% leaky cartons. The new chart is used for a number of weeks without signalling any more special causes. However, Gillian still thinks the overall numbers of leaky cartons is far from acceptable, so she decides to determine if she can discover any common causes. She stratifies the data according to time of day and discovers that more leaky cartons are produced in the morning than in the afternoon. She also finds that most of the leaks are caused within the first hour of production, and on investigation the process engineer notes that the sealer does not operate as efficiently at the start of production because it needs to reach a high operating temperature. Consequently they decide to turn on the sealer 30 minutes before the start of production. After this improvement, Gillian reestablishes the chart with new data from the process and discovers that the percentage of leaky cartons has fallen to just under 2%. Gillian continues monitoring with the chart looking for further opportunities for improvement.

12.12 Chapter Summary

This chapter has covered the construction of different control charts and how the charts are used to monitor business processes. Specifically:

- A control chart is a run chart with a centre line at the arithmetic mean and upper and lower control limits at *three sigma* above and below the centre line, respectively.
- Variables charts for interval data come in two varieties:
  - The individuals chart, where the data are not in subgroups and *sigma* is equal to the standard deviation of the data.
  - The R and X-bar charts, where the data are in subgroups and *sigma* is based on the mean range of the subgroups.
- Attribute charts for proportions and counts are of two types:
  - The p-chart, which is used for proportions and where *sigma* is obtained from the standard deviation of a proportion.
  - The c-chart, which is used for count data and where *sigma* is obtained from the standard deviation of a count.
- For all charts it is important to eliminate any data that have identifiable special causes when establishing control charts.
- Variables charts, based on interval data, are usually more informative that attribute charts.
- Charts can be made more or less sensitive by increasing or decreasing the frequency of sampling.
- A chart that is too sensitive will signal special causes too often, resulting in many process stoppages for which there may be inadequate resources to investigate all the special causes.

12.13 Exercises

1. G&S Whiteware is having difficulties in machining spindles for its new Soft Suzie washing machine. Its decides to set up control charts for the width of the spindle by sampling four spindles every hour, for the next two days or so. The widths of the spindles, in millimetres, are given in Figure 12.15, together with subgroup means, ranges, and overall means (in F27 and G27).
   a. Calculate the control limits, and set up control charts.
   b. What conclusions do you draw from your control charts?

2. Consider the Nuva Plastics data given in the Excel spreadsheet in Figure 12.5. No special causes showed up on the R chart in Figure 12.6 or the X-bar chart in Figure 12.7. The process that produced these data is, therefore, a stable process. Or is it?
   a. Look closely at the data in each subgroup. What do you notice?
   b. Why do you think it happened?
   c. What do you suggest as a remedy?

3. Helen Cassells was reviewing the figures for last year's video camera sales. One of the models, the Z20A, had been selling well at the beginning of the year but now looks to be in decline. She called Neville Watson, brand manager in charge of this model's sales, to see what had been happening.

	A	B	C	D	E	F	G	H
	Subgroup	Item1	Item2	Item3	Item4	Mean	Range	Notes
1	1	10.3	11.5	9.8	12.3	11.0	2.5	
2	2	9.9	8.7	10.1	11.2	10.0	2.5	GH absent
3	3	13.2	12.9	13.4	13.3	13.2	0.5	
4	4	12.6	12.3	11.9	12.2	12.3	0.7	
5	5	10.0	12.8	12.8	12.4	12.0	2.8	
6	6	10.5	13.1	13.2	13.1	12.5	2.7	
7	7	12.3	11.1	13.1	15.9	13.1	4.8	New operator
8	8	11.9	10.1	13.0	13.4	12.1	3.3	
9	9	11.8	9.9	12.9	12.2	11.7	3.0	
10	10	11.7	9.8	13.3	11.1	11.5	3.5	
11	11	10.3	10.6	10.5	13.8	11.3	3.5	
12	12	10.2	12.4	14.2	12.2	12.3	4.0	
13	13	12.6	13.9	10.7	13.1	12.6	3.2	
14	14	10.8	13.2	14.2	12.0	12.6	3.4	
15	15	9.8	11.1	12.2	13.0	11.5	3.2	
16	16	12.3	12.8	13.2	13.5	13.0	1.2	
17	17	11.0	13.2	14.3	12.6	12.8	3.3	
18	18	9.0	14.0	12.3	12.3	11.9	5.0	Oil ran out
19	19	15.0	12.3	13.4	12.9	13.4	2.7	
20	20	12.0	12.5	12.2	14.3	12.8	2.3	
21	21	12.5	12.4	12.5	12.6	12.5	0.2	
22	22	9.5	12.6	13.2	13.2	12.1	3.7	
23	23	11.3	11.6	11.9	12.3	11.8	1.0	
24	24	14.3	12.4	11.6	12.5	12.7	2.7	
25	25	9.3	12.7	10.4	12.3	11.2	3.4	
26						12.14	2.76	

Sheet1 / Sheet2 / Sheet3

Figure 12.15 Spindles data.

"Neville, have you seen the latest figures on the Z20A? Since you employed that new salesperson at the end of October, there's been a downturn in sales."

"Yes, Helen, I'd noticed that too, but I think we should look at the monthly figures for the last two years before we jump to any conclusions."

"OK, Neville, but can you get back to me about this before the end of the month with a recommendation about what we should do to stop a further decline in the sales of the Z20A."

Neville collected the sales figures for the last two years, which are given in Figure 12.16. The mean value of sales is 17 (thousand dollars) with a standard deviation of 1.10 (thousand dollars).

a. Calculate the control limits for these sales, and draw them on a control chart.

b. What conclusions do you draw from the control chart?

c. What should Neville do next?

Month 2003	Jan	Feb	Mar	Apr	May	Jun	Jul	Aug	Sep	Oct	Nov	Dec
Sales (in thousands of dollars)	15.4	15.6	17.2	16.8	16.3	16.6	17.3	18.0	17.2	16.5	17.5	19.2
Month 2004	Jan	Feb	Mar	Apr	May	Jun	Jul	Aug	Sep	Oct	Nov	Dec
Sales (in thousands of dollars)	18.2	17.3	17.4	16.3	15.4	17.2	17.3	15.5	18.9	19.1	16.4	16.1

Figure 12.16 Sales of the Z20A model.

4. In question 1 in Section 11.11 a policy to control costs was discussed. In order to examine this policy more closely, the cost variances, expressed as a percentage over or under budget, on the last 20 construction projects were obtained. They are given in Figure 12.17. The mean and standard deviation are 1.425% and 10.17%, respectively.
    a. Calculate the upper and lower control limits, and plot them on a control chart.
    b. Review your answer to Question 1 in Section 11.11: "What are your reactions to this policy?"

5. Elaine is a line supervisor responsible for filling cans with milk powder at a local dairy factory. She has been having problems getting the correct amount of powder in the cans. She has to ensure that she does not underfill the 2 kg cans, but at the same time the company does not want to give too much powder away. There are two settings on the can filling machine, high and low. Elaine is collecting data on this process, and when she gets a result that she regards as too high she resets the machine to its low setting, but if it is too low she resets it to its high setting. Figure 12.18 gives the control chart constructed from the data Elaine collected.
    a. Is there any evidence of special causes in this control chart?
    b. What do you think has happened here?

6. A run chart for 12 red bead experiments, consisting of 300 results, is given in Figure 12.19. The mean number of red beads is 10.1, with a standard deviation of 3.1.
    a. Calculate the control limits, and plot them on the run chart in Figure 12.19.
    b. What does a result outside these limits mean?
    c. What proportion of the results falls outside your control limits?
    d. Assuming that the results come from a normal distribution with mean 10.1 and standard deviation 3.1, what proportion of red beads would be expected to fall outside these limits over a large number of experiments?

Project	A	B	C	D	E	F	G	H	I	J
Cost Variance	-12.5	0.1	-6.1	10.1	10.1	9.5	-12.5	1.3	1.3	32.0
Project	K	L	M	N	O	P	Q	R	S	T
Cost Variance	-0.7	7.8	0.7	-5.7	3.5	-9.1	-5.2	9.1	0.7	-5.9

Figure 12.17 Project cost variances (%).

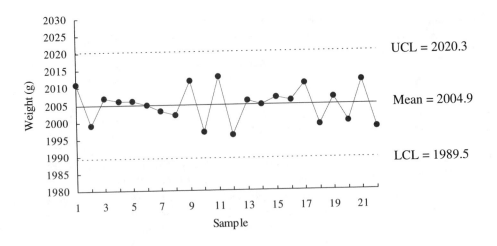

Figure 12.18   Weight of cans of milk powder.

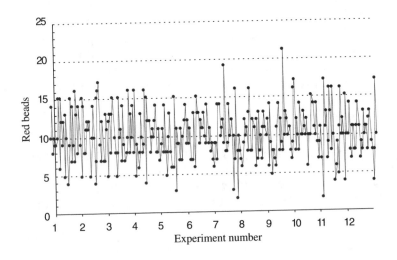

Figure 12.19   Results of 12 red beads experiments.

7. Azaka Rims produces rims for bicycle wheels. Leslie Kerr, the sales manager, is evaluating the performance of the sales department. She suspects that some sales staff are not performing as well as they could, and it might be time for another meeting to review selling methods. However, she recalls from her studies at university that she should look at data, so she collects data on aggregate sales for the last 23 days, which are given in Figure 12.20.
    a. Draw a run chart and set up the control chart for this set of data.
    b. What conclusions can you draw from this control chart?

8. Write down three likely causes for an R chart to go out of control.

9. The Northmost Packaging Company supplies milk cartons to Northmost Dairy Ltd. Recently John Nikau, the dispatch manager for Northmost Packaging, has had several calls from Helen Carter, the quality control manager at the dairy, about leaking cartons. John organised a team to investigate the

Day	Rims	Notes	Day	Rims	Notes
1	110		13	170	
2	142		14	161	
3	131		15	69	Jim W overseas
4	159		16	143	
5	187		17	154	
6	189		18	129	
7	242		19	107	
8	174		20	158	
9	153		21	192	Sales conference
10	166		22	178	
11	140		23	156	
12	152		Total	3562	

Figure 12.20   Sales of bicycle wheel rims.

problem, and the team decided to set up a control chart to monitor its improvement effort. The team estimated that about 10%–12% of cartons were leaking. It collected samples of 50 cartons and counted the number of leaking cartons, to give the data in Figure 12.21.
a. Why did the team choose samples as large as 50?
b. Calculate the control limits for this set of data.
c. Plot the data and control limits on a run chart.
d. What conclusions do you draw from your control chart?
e. What do you think the team should do next?

Sample number	Leaking cartons	Proportion	Sample number	Leaking cartons	Proportion
1	5	0.10	13	5	0.10
2	7	0.14	14	1	0.02
3	2	0.04	15	1	0.02
4	0	0.00	16	2	0.04
5	12	0.24	17	4	0.08
6	3	0.06	18	0	0.00
7	7	0.14	19	7	0.14
8	7	0.14	20	5	0.10
9	6	0.12	21	8	0.16
10	8	0.16	22	7	0.14
11	12	0.24	23	4	0.08
12	3	0.06	24	1	0.02

Figure 12.21   Leaking cartons at Northmost Packaging.

Day	Number	Day	Number	Day	Number	Day	Number
1	35	7	50	13	34	19	55
2	42	8	75	14	40	20	60
3	57	9	43	15	47	21	52
4	38	10	27	16	48	22	51
5	40	11	49	17	56	23	63
6	44	12	36	18	49	24	49

Figure 12.22   Defects on washing machines.

10. To set up a control chart for surface defects, a quality manager takes a sample of 20 washing machines from each day's production over a four-week period (i.e., 24 days). The numbers of surface defects (e.g., scratches, paint blemishes) each day are shown in Figure 12.22.
    a. Calculate the average number of defects per day and the average per machine.
    b. What is the most appropriate form of control chart for these data?
    c. Calculate the control limits, and draw a control chart for the data.
    d. What do you conclude from your control chart? What would you do next?
11. Ray Higgins is product manager at Murphy Electric in charge of worldwide sales of the Z78A fax machine. He has been reviewing the sales figures for 2003 and 2004 only to discover that the last three months (October to December) seem to show a decline in what had been an upward trend, as shown in the run chart in Figure 12.23. He decides to look at the sales figures by salesperson and discovers a general fall in sales for the last three months without any obvious patterns. Consequently he sets up a meeting of all the sales staff to give a pep talk.
    a. What are some of the consequences of Ray's actions, and what should Ray have done?

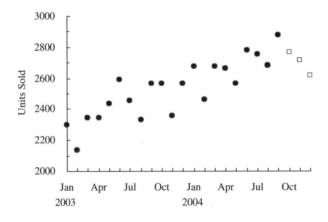

Figure 12.23   Sales of fax machines.

| 2003 | | | 2004 | | |
Month	Units Sold	Residuals	Month	Units Sold	Residuals
Jan	2300	26	Jan	2674	94
Feb	2134	-166	Feb	2459	-147
Mar	2345	20	Mar	2678	47
Apr	2342	-9	Apr	2665	8
May	2432	56	May	2564	-118
Jun	2589	187	Jun	2778	70
Jul	2456	29	Jul	2756	23
Aug	2333	-120	Aug	2679	-80
Sep	2567	89	Sep	2875	91
Oct	2564	60			
Nov	2356	-173			
Dec	2568	13			

Figure 12.24 Regression residuals for fax machine sales.

Ray gets shifted to another job and Raylene Boyle is promoted to product manager. Having studied statistics at university, Raylene decides to analyse Ray's sales figures given in Figure 12.23. First she calculates the regression line to be

$$\text{Sales} = 2249 + 25.5 \times \text{Month}$$

where month is numbered from 1 to 21. She then calculates the residuals from this line, which are shown in Figure 12.24.

b. The standard error of the residuals is 100.1. Calculate the upper and lower control limits, and draw them on the run chart, along with the regression line.

c. What conclusions can you draw from your control chart?

12. George Wilson runs a debt collection agency. The last several months' figures show an increase in the total amount of money collected. Recently the company has renegotiated some of its contracts for collection by placing some people on commission. George decides that it would be a good idea to put more people onto the new contract, and he arranges for a meeting of all union representatives to start negotiations.

a. What could be some of the consequences of George's action?

b. What should George have done?

c. The data on monthly receipts are shown in Figure 12.25. The standard deviation of monthly receipts is 10.515. Calculate the control limits, and set up the control chart for these data.

d. What conclusions can you draw from your control chart?

e. What would you do next?

Month	Receipts (thousands of dollars)	Month	Receipts (thousands of dollars)	Month	Receipts (thousands of dollars)	Month	Receipts (thousands of dollars)	Month	Receipts (thousands of dollars)
1	127.8	6	119.1	11	118.7	16	127.1	21	110.4
2	109.5	7	143.7	12	133.1	17	129.5	22	135.3
3	124.0	8	115.2	13	125.0	18	122.4	23	137.7
4	114.0	9	125.3	14	129.4	19	123.6	24	140.2
5	141.7	10	135.6	15	121.7	20	108.2	25	142.3
								Total	3160.5

Figure 12.25   Monthly debt collection receipts.

| 58 | 51 | 48 | 46 | 50 | 47 | 68 | 59 | 53 | 57 | 43 | 44 | 32 |
| 45 | 50 | 40 | 50 | 67 | 44 | 45 | 51 | 69 | 34 | 49 | 50 |

Figure 12.26   Toffee content of hoki-poki ice cream.

13. Jacques Henri is in charge of the production process that produces hoki-poki ice-cream at Bit-Bot Ltd. Ideally, each 1 kg carton of ice cream should contain 50 spherical pieces of toffee, but it is acceptable to customers as long as it has between 45 and 55 pieces. Recently they have had complaints from customers of not enough toffee in their hoki-poki, and Jacques has decided to up the average amount of toffee to 60 pieces.
    a. What could be some of the consequences of Jacques' action?
    b. Aldi Hyde, the line supervisor for the process that produces hoki-poki, decides to collect some data. She takes 25 packs of ice cream and counts the number of toffee pieces in each, yielding the data in Figure 12.26. Which type of control chart is most appropriate for this set of data?
    c. The mean of the data in Figure 12.26 is 50 and the standard deviation is 9.26. Set up the control chart for this process.
    d. What does the control chart tell you? Review Jacques' initial action.

Chapter 13

Improvement Strategies

13.1 Introduction

As we have remarked earlier, statistical process control is part of an overall plan to reduce variation and improve business processes. It has been found to be an effective way to improve product and service quality. In the previous chapter we discussed different types of control charts and showed how these charts can help us separate out special and common causes of variation. Once a chart has been established it can be used to monitor the process in the future.

However, the process may not be accomplishing what it should or could be achieving. Special causes may be present, and these will have to be identified and eliminated if the process is to be bought under statistical control. Even a process in statistical control may not be satisfactory because it may be producing items that are off-target or with too much variability. Although the process is in control, it may not be meeting the process specifications. In other words, the process may not be capable of satisfying the needs of the customer.

It is only by improving the process that it can be bought under control, meet specifications, and be capable of satisfying customers' needs. How this can be achieved is considered in this chapter.

13.2 Improvement Strategies

We have seen that control charts can be used to identify whether variation arises from common or special causes. In Chapter 11 we pointed out the fundamentally different nature of special causes of variation from that of common causes. Hence, if we are seeking to improve a process, different improvement strategies are required for processes that exhibit common cause variation from those where special causes are involved.

Sometimes special causes lead to desirable outcomes, for example, increased sales as a result of a TV advertising campaign. However, if the special causes have undesirable consequences, then they need to be eliminated. This means finding out what was different at the time they occurred in order to prevent them from reappearing. Eliminating special causes is the first priority. They are usually easier to detect and fix, and they tend to mask the common cause variation.

Remember the 85/15 rule, though, where most problems are due to *common causes*. Common cause problems are harder to detect, involve changes to the system, and invariably require management action. However, once we have succeeded in dealing with the special causes, further improvements can be made only by developing strategies to reduce common cause variation.

✎ Which Strategy?

Fixing each type of cause requires a different managerial strategy. We use control charts to identify whether we use a common cause strategy or a special cause strategy. Consider the control chart in Figure 13.1, which has been established for TV sales data on successive Saturdays. The chart is now being used to monitor sales from Saturday to Saturday.

> **Q1.** Which improvement strategy would you use if TV sales were $47,000 on the 26th Saturday?
>
> **Q2.** What if sales were $31,000?

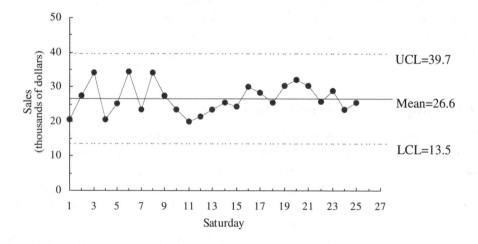

Figure 13.1   TV sales.

13.3 Special Cause Strategies

> Steve arrived in the office that morning to discover that his company had not delivered Kaplan's order on time. For the last three years, the monthly order had always been on time. What had changed?

Here we probably have a special cause, with an opportunity to improve delivery performance. Not all special causes are so obvious, and we need control charts to spot them. But if we miss the opportunity of finding out what is different now the problem is likely to recur.

> Steve phoned Kaplan and spoke to their purchasing manager, and explained that the order would be a day late. He then set about looking into the problem. Upon investigation Steve discovered that Kaplan's order had been late because the company had to re-run the whole batch of boxes again because somebody had printed Kaplan's new telephone number upside down on the outside of the box. Further investigation revealed that this had happened on several orders, and was caused by a recent change in local telephone numbers.

A common mistake is to fix the problem by adding extra steps to the process, in this case, perhaps by adding an inspection stage before the batch of boxes is printed. Instead they redesigned all printing "slugs" by changing them from clean oblong shapes to oblong shapes with a protruding notch, as depicted in Figure 13.2. In this way, they could not be put into the printing machine upside down.

The example illustrates the important stages in dealing with special cause problems:

1. Get timely data so that special causes are signalled quickly. A speedy reaction to a special cause problem is very important. We must therefore get data immediately so that we can investigate the problem when the trail is hot. For this reason it is best to place the control chart near the process.
2. Put in place an immediate remedy to contain any damage. This is "firefighting," but it is necessary, for example, to take action to placate an angry customer. Firefighting is not acceptable if we were to stop there.
3. Find out what was different about the special cause. Has something changed? Is there perhaps a new operator, a new competitor, or a change in procedure?

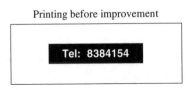

Printing before improvement

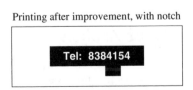

Printing after improvement, with notch

Figure 13.2 Redesigned printing slugs.

4. Develop a long term fix. We should have learned something in the search for the cause. What can we do to stop its recurrence or, if the results were good, to maintain this level of performance? This could mean changing some higher-level system, which often brings benefits beyond the immediate problem. In Steve's case he changed the printing procedure, and this prevented upside down printing on all orders, not just Kaplan's and not just telephone numbers.

Q3. *You have been getting A's all year in assignments and tests. Suddenly you get a C. How would you apply special cause strategies to this problem?*

13.4 Common Cause Strategies

A process with only common cause variation is stable and predictable. However, it may not be meeting customer needs. For example, the beads process is a stable and predictable process, but we are producing too many red beads (defects).

Improving a process with only common causes of variation present needs a different approach than that of eliminating special causes. Trying to find out what was different between this data value and the others, as we would with a special cause strategy, will not necessarily yield any answers. All the data needs to be examined. Common causes are usually difficult to discover. They do not suddenly present themselves like special causes. They are there all the time and need to be found by analysing all the data. *With common causes, quick fixes rarely work.* What we need are strategies that help us understand the workings of hidden causes present in the system all the time. The aim is to localise common cause variation by pinpointing it at its source. Once we know that, we can develop measures to counteract the common cause variation by making *fundamental changes* in the system.

There are three major strategies we can adopt: root cause analysis, stratification, and experimentation. We shall now consider each of them in turn.

♘ *Root Cause Analysis*

We must be able to identify the major potential causes that create some of the variation. We need to dig down to find the root causes. This means not just accepting the first explanation of what is happening, but continuing to question and probe until the real causes have been determined. The cause and effect, or fishbone, diagram is a useful tool for displaying the potential causes.

📈 Debt Recovery Project

In the case study in Section 2.6, root cause analysis was used to reduce the level of unrecovered debt. After mapping out the process, a cause and effect diagram was used to list all the potential causes of overdue accounts. The team identified the most probable causes and collected data to verify which were likely to be the root causes. It then proposed solutions to reduce the variation and the overall level of outstanding debt.

🔌 Stratification

In Chapter 3 we saw how stratification could be used to uncover previously unknown features by disaggregating data into meaningful subgroups. In a similar way, we can look at data that have been collected from a process by sorting it into groups or categories based on different factors. In this way we can seek patterns in the way the data cluster or do not cluster. Such patterns may well reveal important common causes of variability.

📈 Customer Deliveries

A company assembling and selling desktop and notebook computers sells these exclusively through the Internet. Following the receipt of an order, the particular model and specification ordered is scheduled for production, and the customer is given a firm date when the computer will be delivered, by courier, to the customer's address. In recent months, delivery dates have been falling behind schedule, with some customers not receiving their order until two or three weeks after the promised date. This is causing increasing concern because the company has built its reputation on product reliability and prompt delivery. To address the problem, the sales manager decides to collect and analyse data from each stage of the production scheduling and delivery processes, for this may help discover common causes of late delivery.

He decides to look at the data first by product type and customer location. This reveals significantly more late deliveries for notebook versus desktop models. Because the notebook models involve more bought-in parts than desktop machines, he now decides to examine the raw material ordering and delivery data to see if he can find evidence of delays in notebook production caused by nonavailability of parts. He finds that the company that supplies the carrying cases for the notebooks has become increasingly unreliable in recent months, resulting in delays to the final packaging of notebooks. He decides to renegotiate delivery schedules with the supplier, with the threat that they will switch to a new supplier if performance does not improve.

🔌 Experimentation

When it is difficult to discover the reasons for common cause variation from data that already exist or from data collected from the process running under normal conditions, we may decide to carry out an experiment on the process. This involves making carefully planned changes to a process, analysing the results, and observing any effects. Experimentation is usually expensive and can often disrupt the process, but it does allow us to find causal relationships between inputs and outputs.

Many issues and concepts are involved in planning and carrying out useful industrial experiments. We shall demonstrate some of these with the helicopter experiment in a case study in Section 13.7. For now, to clarify some ideas, consider the following two experiments.

🏭 Cardboard Manufacture

A machine that produces corrugated cardboard (a corrugator) was improved using a planned experiment. One of the qualities important for corrugated cardboard is thickness. It is well known that variation in the thickness of the board depends on numerous machine settings, such as temperature, pressures of different rollers, the speed of processing, and the type of glue. The process is so complex that data collected on board thickness had yielded no solution to the problem of excessive variation in thickness, so an experiment was undertaken. This involved setting the corrugator at different temperatures, roller pressures, and speeds in a planned way to discover the causes of the variation. As a result of the experiment the amount of variation in the thickness of the corrugated cardboard was reduced.

🏭 Plastics Moulding

The operations manager of a plastics company was concerned by the variability in the shrinkage of product obtained from an injection moulding machine. She decided to carry out a planned experiment to try to discover the causes of the excessive variation. Her experiment involved setting the levels of a number of variables she considered likely to affect output. These included cycle time, mould temperature, holding pressure, injection speed, holding time, and gate size. Carrying out the experiment led to a reduction in the amount of shrinkage in the product.

13.5 Specifications

In the manufacturing and process industries, specifications are set for the most critical characteristics, and failure to meet these specifications is one important reason for seeking to improve a process. For example, a dairy company may aim to fill 2 kg cans of milk powder with no more than 2.01 kg of powder. The specification limits are, therefore, 2 kg to 2.01 kg. Any cans containing less than the lower specification limit (LSL) would violate their legal requirement, while filling with more than the upper specification limit (USL) would mean the dairy company was giving away too much powder.

In another example, suppose we are making electric motors. Our requirements are that the main shaft be round and 25 mm in diameter. The target is 25 mm, but not all shafts will be 25 mm in diameter because some variability is inevitable. As long as they are not too much off target, they can be used. Specification limits might therefore be set at 25 mm $\pm$ 0.1 mm.

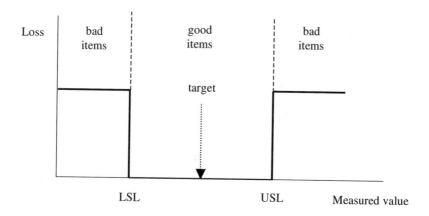

Figure 13.3   Conventional view of specifications.

The same concepts apply with business processes and in the service industries. A freight company might set specification limits on the time it takes to deliver a parcel. A delivery that takes longer than a certain time (USL) may involve some penalty. For example, one parcel delivery company guarantees delivery within 24 hours, or you get your money back. A telephone company may say they are providing a good service if your telephone fault is rectified within eight hours. The finance department may be content if it pays travelling expenses within two weeks. An airline may have a specification that all baggage must reach the carousel within 20 minutes of the plane's arrival.

The assumption is that if all parts of a product conform to specifications, then the entire product will be of good quality. The conventional view of specifications is as an all or nothing situation. Items are either good or bad. The baggage either arrives within the specified time limits or it does not. This view is depicted in Figure 13.3.

The conventional view that conformance to specification equates to good quality is simplistic. It reveals little interest in whether the target has been met exactly. Variability *within* specification is, for the most part, ignored. A further drawback is the problem known as *tolerance stack-up*. When two or more parts are to be matched together, the fit might be poor if one part is at the lower end of its specification and the matching part at the upper end. As a result the link between them may wear more quickly than one made from parts whose targets have been matched exactly. Repairing products at a later time can also cause problems in finding matching parts.

A more realistic view is given by the loss function depicted in Figure 13.4. This way of looking at specifications is due to the Japanese engineer Genichi Taguchi. The ideal is at the target, and any movement away from it causes increasing losses. Losses near the target value may be negligible, but as you move further away they become greater. The loss function is *not* a step function. We have ever increasing losses with departure in either direction from the target value. Neither this function nor the one in Figure 13.5 has to be symmetric. It may even be one-sided, as with the telephone company, where the ideal is to fix the fault immediately.

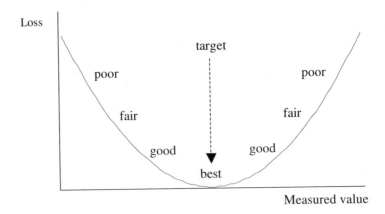

Figure 13.4   Taguchi's loss function.

Consider the following situation, where we have the choice between two suppliers, A and B, for some component. The distribution of the critical dimension of the component is shown for both suppliers in Figure 13.5 together with the specification limits.

Q4. *Both suppliers have all their components within the specification limits. Which supplier would you choose, and why?*

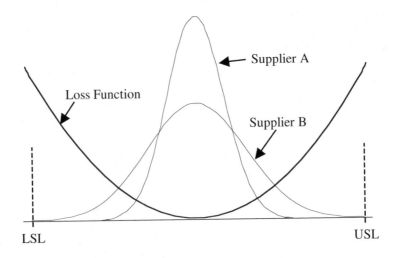

Figure 13.5   Component distributions for suppliers A and B.

✤ *Specifications and Control Limits*

We have now considered two types of limits: *control limits* and *specification limits*. They are often confused in practice; but it is important not to do so.

Control limits are about the performance of the process. They tell us what the process is doing right now. They are calculated from the data collected on the process. They are the *voice of the process*. We use control charts to help us distinguish between special and common causes of variation. The aim is to eliminate the special causes so that we have a stable and predictable process. It does not necessarily mean that we are happy with the process or that it meets our needs.

Specification limits reflect what we desire from the process. They are limits ultimately set by the customer and are the *voice of the customer*. They are chosen by the process engineer to reflect what the customer wants. For example, if the customer wants the door on the refrigerator to close properly and tightly each time, specification limits are set on factors such as the size of the door, the hinge mechanism, and the frame dimensions in order to achieve that requirement. They are not determined by the data.

It is important *not* to draw specification limits on a control chart. It is a dangerous practice that can badly mislead you.

Figure 13.6 gives the X-bar chart for the Nuva Plastics example of Section 12.6. Suppose the specification limits are 113 ±0.5 grams. Draw them on the X-bar chart (never do this again!).

> **Q5.** *Are all the data within specification limits?*
> **Q6.** *In what way can you be misled by what you have done?*

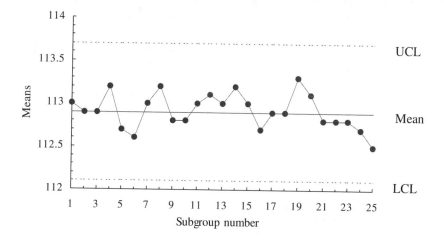

Figure 13.6 Nuva Plastics X-bar chart.

You should see in your answers to these questions that you can be seriously misled by plotting specification limits on an X-bar chart. Although we would not have the same problem with the individuals chart, p-chart, or c-chart, we still suggest that you not draw specification limits on any control chart. This is because control charts should be used only for deciding whether a process is in statistical control and for monitoring and improving the process. Instead it is better to draw specification limits on a stem and leaf plot if you wish to look at whether the outputs from a process are meeting specifications.

13.6 Process Capability

Suppose we have a process that is in statistical control. Does this mean the process meets customer needs, as set by the specification limits? Is our entire product within the specification limits, or do we have to reject product because it falls outside these limits. In other words, is our process *capable*?

A process can either be in statistical control or not, and the product can either meet the specifications or not. Hence, when considering the capability of a process, four states are to be considered. Only in the first of these states is the process capable.

⚡ *Ideal State*

The ideal state is when the process is in statistical control and all of the output is within the specification limits. It means that all the output is suitable for the intended purpose, and as long as it stays in control the process will produce consistent output hour after hour, day after day, week after week. The aim is to stay in this state. Control charts are used to signal problems before they are severe enough for the process to become incapable. Even in this state there may be reasons for further improving the process. Lower costs and greater productivity can be achieved from even more uniform output, as shown by Taguchi's loss function in Figure 13.4.

⚡ *Threshold State*

If the process is stable but some of the output is outside the specification limits, then we are in what is called the *threshold state*. A process in this state is judged incapable. For a process that is stable but incapable, the process may not be centred on target, or may be too variable, or may be a mixture of each.

Different remedies will be needed to resolve the two different sources of incapability. Specifically:

- The process can often be centred on the target by changing the operating parameters. Experimentation often reveals how to make these adjustments.
- The reduction of excessive variability requires removal of some of the common causes of variation and is usually more difficult.

Control charts should be used to ensure that the process is maintained in statistical control and also as a device to monitor improvement.

৬ *Brink of Chaos*

We are in the brink of chaos state if the process is unstable but all of the output is within specification. But what is wrong with that? Surely customer needs are being met.

It is not possible to say anything about the capability of a process that is unstable. Since special causes are present, we no longer have a predictable process. Although in this state the product right now is all within specification, it is likely to wander off and produce output that is out of specification. The problem is that special causes are determining what the process is producing. Quality and conformance can change at random, thus leading to chaos. It is necessary to eliminate these special causes.

৬ *Chaos*

Chaos results if the process is out of control and is producing nonconforming products. Again the process is dominated by special causes. The only way out of this state, into the ideal and threshold states, is first to eliminate special causes. Control charts are again invaluable as a tool to assess whether special causes are being eliminated.

📉 *Airport Immigration*

An international airport has to maintain processing standards set by the government by ensuring that 95% of passengers arriving on a flight are processed through immigration within 45 minutes of landing. Data collected for all flights landing over two days are given in the stem and leaf plot in Figure 13.7. Each observation gives the time (to the nearest minute) for 95% of passengers from a flight to be cleared through immigration. The data were obtained from the computer system used to record arriving passengers. Mark in the specification limit on the stem and leaf plot.

Q7. *What percentage of flights is not meeting the standard?*
Q8. *What type of control chart is appropriate for these data?*

The control chart is given in Figure 13.8.

Frequency	Stem	Leaf
1	2	2
0	2	
1	2	7
4	2	8899
9	3	000001111
7	3	2233333
12	3	444444455555
16	3	6666666666777777
11	3	88888889999
15	4	000000000011111
5	4	22223
5	4	44445
1	4	6
0	4	
3	5	001

Unit	10	1
Stem increment		2

Figure 13.7   Immigration processing times.

Q9. What is the capability state of this process?
Q10. What is the first step to take in improving the process?

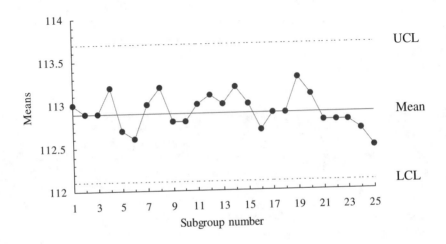

Figure 13.8   Control chart for immigration processing times.

13.7 Case Study: The Helicopter Experiment

The aim of the helicopter experiment is to show you the power of statistical experimentation in uncovering hidden factors that influence the performance of a process. In this demonstration we look at how we can improve the design of a paper helicopter. All of the ideas extend to real industrial experiments, such as that of the corrugator or the injection-moulding machine in Section 13.4.

A picture of a helicopter is shown in Figure 13.9. It is made by cutting and folding a piece of stiff paper or card into a shape comprising a vertical shaft with two horizontal rotors. To give added stability, the shaft may be stiffened by folding together multiple layers or by adding some form of weight to the bottom, such as one or more paper clips.

Q11. How will we measure the performance of our paper helicopter? How will we know if we have made a good one?

Q12. We can vary many things in the design of the helicopter. Make a list of the factors you think are likely to influence the helicopter's performance.

Q13. Each of the factors listed can take many different values, or levels. What levels should we use for each?

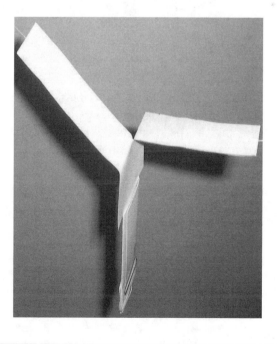

Figure 13.9   A paper helicopter.

Factors	Levels	
1 Type of paper	thick	thin
2 Rotor width	2 cm	4 cm
3 Rotor length	10 cm	12 cm
4 Weight	no clips	2 clips
5 Type of shaft	fold	cut
6 Fold on bottom	no	yes

Figure 13.10   Helicopter factors and levels.

✍ Designing the Helicopter

Suppose we have chosen the factors and levels shown in Figure 13.10.

Q14. How many different paper helicopters can we make?
*Q15. We could make all the helicopters, but it would take a lot of time to carry
out the experiment. It would also be expensive and time consuming. How
else could we run the experiment?*

✍ Helicopter Experiment Design

The experiment is run with eight different helicopters involving six factors, each
at two levels. In Figure 13.11 the run number is given in the first column and
the *design* of the helicopter in the next six columns. The last two columns will
be used to record the results of the experiment.

*Q16. What patterns can you see in the levels of the factors in the design in
Figure 13.11? (Hint: Look carefully at each column and each pair of
columns.)*
*Q17. Should we run the experiment in the order given in Figure 13.11? If not,
how do we decide the run order?*

Run	Type of paper	Rotor width	Rotor length	Weight (Paper clips)	Type of shaft	Fold on bottom	Time	Quality
1	thick	2 cm	10 cm	0	cut	yes		
2	thin	2 cm	10 cm	2	fold	yes		
3	thick	4 cm	10 cm	2	cut	no		
4	thin	4 cm	10 cm	0	fold	no		
5	thick	2 cm	12 cm	0	fold	no		
6	thin	2 cm	12 cm	2	cut	no		
7	thick	4 cm	12 cm	2	fold	yes		
8	thin	4 cm	12 cm	0	cut	yes		

Figure 13.11   Design of the helicopter experiment.

Running the Experiment

To run the helicopter experiment you first need to assemble the eight helicopters in Figure 13.11. Templates for these helicopters are given in *Helicopter.xls* on the CD-ROM. The second worksheet, *Instructions,* tells you how to construct the helicopters, while the templates themselves are in worksheets *Unit 1* through *Unit 8.* Once they have been constructed you can run the experiment by dropping each helicopter in turn according to the run order you determined in your answer to Q17.

When running the experiment, try to drop the helicopter from as high as possible. It is best to hold the helicopter between the first finger and thumb, just under the rotor arms, so that the rotor arms are supported ready for flight. Ensure that the same method is used for each helicopter. The helicopter is dropped when the timekeeper gives the signal. A stopwatch is used to determine the time taken for the helicopter to hit the floor. It is also useful to have a measure of the quality of flight, and this can be based on a 6-point subjective scale. A beautifully spinning, slowly descending helicopter is a 5, and one that flops miserably is a 0. Carry out a couple of trial runs to familiarise the partici-pants with the process. In particular, it gives some benchmarks for using the 0–5 scale.

Plotting the Data

Two responses are obtained when we drop each helicopter, and these are recorded in the last two columns in Figure 13.11. If you did not run the experiment, copy the data given in upcoming Figure 13.17 into Figure 13.11. The first response is the time (in seconds) for the helicopter to land on the floor. The second is a measure of the quality of the flight, on a scale from 0 to 5.

Having collected the data, we can calculate the total scores for each factor level. For example, the time and quality score for a 10 cm rotor length are the totals of

Totals	Type of paper		Rotor width		Rotor length		Weight (No. of clips)		Type of shaft		Fold on bottom	
	thick	thin	2 cm	4 cm	10 cm	12 cm	0	2	fold	cut	yes	no
Time												
Quality												

Figure 13.12   Total scores for the helicopter experiment.

rows 1–4 in Figure 13.11, whereas the values for two paper clips are obtained by summing rows 2, 3, 6, and 7. These total scores can be entered into Figure 13.12.

The total scores for time can now be plotted in Figure 13.13 and the total scores for quality in Figure 13.14. The scores for the two levels for each factor are then joined to show the change caused by varying that factor.

As an alternative to using Figures 13.11 through 13.14 to record the results of your experiment, you can use the first worksheet, *Experiment,* in the spreadsheet *Helicopter.xls.* As the results are entered into the first table, corresponding to Figure 13.11, the other table (Figure 13.12) and the two graphs (Figures 13.13 and 13.14) are automatically constructed. Once all the results of the experiment have been entered, you can then click on *View Time Graph* and *View Quality Graph* to see the graphs.

⚡ Finding the Best Helicopter

The plots in Figures 13.13 and 13.14 will help you determine which factors might influence the choice of the best helicopter. An important factor will have

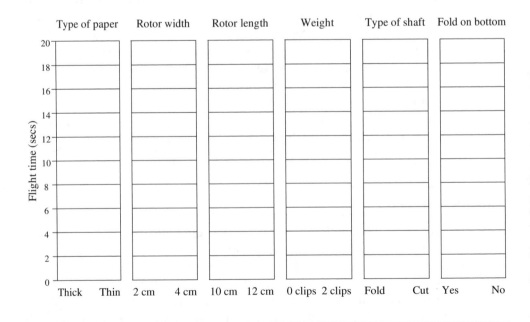

Figure 13.13   Plot of helicopter flight times.

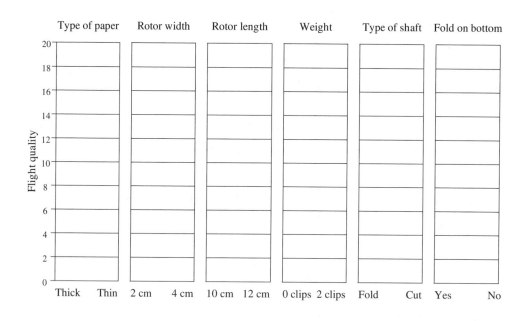

Figure 13.14 Plot of helicopter flight quality.

a large difference between the results at its two levels, depicted by a steep line in your plot from the one level to the other level. Where the choice of level of a factor is not important, the lines will be fairly horizontal.

Q18. *Looking first at the time of flight, which factors have the greatest influence on flight time? Also note the factor level with the longest average flight time.*

Q19. *Repeat this with quality, remembering that the higher the value the better the quality.*

Q20. *Taking both time and quality into account, what levels of each factor would you choose in designing your helicopter? Does this correspond to one of the helicopters used in the experiment?*

Q21. *What should we do next to confirm our results and further improve the helicopter design?*

Q22. *What important concepts have we learned from this experiment that we could apply to real problems?*

Month 2003	Jan	Feb	Mar	Apr	May	Jun	Jul	Aug	Sep	Oct	Nov	Dec
Recycled %	25.6	23.4	25.8	26.2	27.3	24.5	25.2	23.5	27.1	26.1	27.2	28.0
Month 2004	Jan	Feb	Mar	Apr	May	Jun	Jul	Aug	Sep	Oct	Nov	Dec
Recycled %	26.3	25.3	24.7	24.8	25.0	24.6	24.2	24.0	23.9	24.4	23.1	24.2

Figure 13.15   Percentage recovered debts.

13.8 Case Study: Cape Corp Debt Collection Agency

Midi Vaartagen, the manager of Cape Corp Debt Collection Agency, was reviewing progress on the unpaid TV licences contract. Over the last few months it looked as if recovered TV licence debts had dropped. Midi decided not to jump to any immediate conclusions. The last time she thought the recovery of debts had dropped, it turned out to be a temporary downturn, and she had upset quite a few of her staff when she called them in to explain what was happening. So this time she thought she would follow some advice on collecting data that she vaguely remembered from the first year at university when she was studying for her management degree. Figure 13.15 gives data on the percentage of debts they had managed to recover for the TV licensing authority over the last 24 months. The mean of this data is 25.18% and the standard deviation 1.34%.

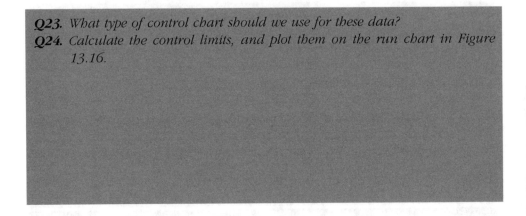

Q23. *What type of control chart should we use for these data?*
Q24. *Calculate the control limits, and plot them on the run chart in Figure 13.16.*

Midi thought about possible actions she could take and came up with the following five options:

1. It looks as if a recent campaign on one of the pirate radio stations may be encouraging people not to pay their TV licences. Contact the minister for telecommunications about this.
2. Dig out further data on TV licence debt collection and categorise them according to employee, department, part of country, etc., to see if any patterns emerge.

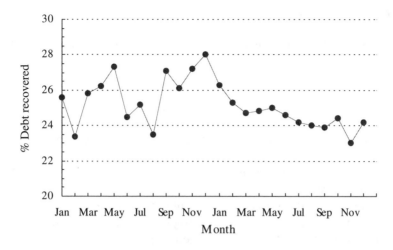

Figure 13.16 TV licence debt collection.

3. Figure out what was different during the last few months compared to other months. Were there more debts to investigate? Were there new employees in the department? Has there been less than usual advertising about the need to buy TV licences?
4. Look at all the debts that had to be followed up in the past few months and categorise the causes of the problems. Look for patterns.
5. Get people working on each major stage in the debt collection process to collect data for several weeks on how many and what kinds of problems arise at each stage.

Midi should first notice that the process is out of control, since the last 10 points are below the mean. It will, therefore, be necessary to use a special cause strategy. Options 2, 4, and 5 are common cause strategies; they could be useful later. Option 1 is a premature solution. Hence, Midi should use option 3, which is a special cause strategy that looks at what has happened in the last few months.

Q25. Suppose Midi decides to use a control chart to monitor the progress of TV licence debt collection. What should be her next step?

13.9 Chapter Summary

This chapter has been concerned with how to use control charts to tell us what type of management strategies should be applied to improve processes. Specifically:

- Special cause strategies aim to reduce variation arising from special causes.
  - Investigate why data from a special cause is different from the rest of the data.
  - Long term "fixes" have to be implemented so that the problem never occurs again.
  - Workers on the process can usually initiate fixes.
  - It is more effective to search for special causes as soon as they occur.
- Common cause strategies are applied to processes in statistical control.
  - Investigate all the data to discover some of the many causes of variation.
  - The three strategies are: root cause analysis, stratification, and experimentation.
- Specification limits (the voice of the customer) are set to reflect customer requirements, and these limits are fixed by the market.
- Control limits (the voice of the process) are calculated from the process data.
- Ideally all items should be at the required target, but some variability is inevitable.
- Taguchi's loss function says that costs for items near to the target may be negligible but that they get larger for items further away from target.
- There are four process capability states, defined by whether a process is in statistical control or not and whether it is meeting specifications or not.
  - *Ideal:* The process is in statistical control and all output meets specification.
  - *Threshold:* The process is in statistical control but some output is out of specification.
  - *Brink of chaos:* The process is not in statistical control and all output meets specification.
  - *Chaos:* The process is not in statistical control and some output is out of specification.
- Process improvement should first work toward the ideal state and then continue reducing variation to produce output that is closer and closer to target.

13.10   Data for the Helicopter Experiment

The results in Figure 13.17 were obtained from an experiment detailed in Figure 13.11.

13.11   Exercises

1. Gertrude Williams is in charge of a printing press that prints customer information and designs onto cardboard boxes. Recently the machine has

Run	Time	Quality
1	2.53	1
2	1.15	3
3	2.54	5
4	1.59	2
5	1.34	1
6	1.99	4
7	1.67	3
8	1.99	2

Figure 13.17 Results for helicopter experiment.

been producing blurred images that have given rise to a large number of customer complaints. A control chart of the data shows no special causes.

a. What improvement strategies should Gertrude consider next?

b. Suppose Gertrude decides to carry out an experiment to investigate the cause of the problem. Engineers tell her that there are several factors that can influence printing clarity: dilution of the inks, ambient temperature, pressure setting of the rollers, speed of the press, type of cardboard, and complexity of the design (number of colours, drawings, etc.). Discuss how you would set up an experiment to identify the important factors. Clearly identify the factors and levels you would use and the way in which you would run the experiment. Write out a set of instructions so that a person could run the experiment in your absence.

2. Consider again Question 3 in Section 12.13. In an attempt to improve sales Neville is looking at the following options:

a. Agree with Helen and suggest further training for the new salesperson. Also suggest that it would be a good idea to set up a sales training course for all of the sales staff, since it is quite a while since they all had a refresher course.

b. Call the new salesperson in for a discussion and find out what he has been doing to sell the Z20A over the past two months, in the hope that you will be able to give him guidance on how to improve.

c. Review the bonus scheme for the sales staff, placing more emphasis on payment by results.

d. Break down the aggregated sales data for the Z20A by area, time of year, customer, and salesperson to look for any patterns.

e. Conclude that the Z20A is reaching the end of its product life cycle and that it is time to develop a replacement model. Set up a market research survey to obtain new information on customer preferences, and report the results to Helen.

Hour	Sample 1	Sample 2	Sample 3	Sample 4	Mean	Range
1	20.74	20.28	23.43	19.55	21.00	3.88
2 *	20.87	21.33	20.96	18.57	20.43	2.76
3	21.45	20.04	22.50	18.89	20.72	3.61
4	28.24	24.11	18.70	22.08	23.28	9.54
5	17.61	23.64	19.52	16.15	19.23	7.49
6	17.65	20.75	19.21	15.99	18.40	4.76
7	20.73	19.65	18.60	20.57	19.89	2.13
8	17.87	20.50	18.79	22.40	19.89	4.53
9	21.50	24.73	20.77	23.69	22.67	3.96
10	22.69	19.38	21.89	16.68	20.16	6.01
11	19.58	21.30	17.78	17.98	19.16	3.52
12	17.92	22.91	22.20	18.86	20.47	4.99
13	22.01	18.63	22.77	23.08	21.62	4.45
14	19.87	19.55	23.28	17.72	20.11	5.56
15	19.91	20.26	21.46	21.98	20.90	2.07
16	17.83	22.59	19.67	23.15	20.81	5.32
17	21.88	22.35	19.94	19.14	20.83	3.21
18	18.92	23.81	21.67	20.39	21.20	4.89
19	20.82	19.22	21.19	16.84	19.52	4.35
20	20.42	19.36	16.34	17.48	18.40	4.08
Totals					408.69	91.11

Figure 13.18 Thickness of plastic laminate.

Which option do you think Neville should choose? Give reasons for your choice and the rejection of the other options.

3. A team of line workers is looking at the thickness of a plastic laminate used to make plastic bags. Every hour they take four samples and measure the thickness (in hundredths of a millimetre) using a laser as the plastic passes over a measuring bench. Figure 13.18 gives the measurements over 20 hours.
   a. Plot X-bar and R charts for these data.
   b. Calculate the control limits, and draw them on the X-bar and R charts.
   c. What sort of strategies would you use to improve this process? List some ways in which you could apply these strategies to this process.

4. Jill Daley works in the dispatch department at Pultron Rods in Hamilton. Part of her job is to ensure that orders are dispatched to meet customer delivery dates. Customers normally order five days before they want the order. Pultron is always on the lookout to improve customer service,

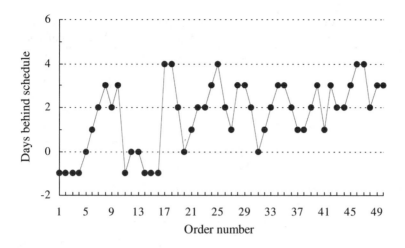

Figure 13.19 Run chart of delivery times.

and Jill is setting up a control chart to help monitor delivery dates. She intends to collect data on 50 delivery times, recording the number of days ahead or behind the requested delivery date that customers receive their orders.

a. Write down two reasons why these data might be difficult to obtain.

b. Jill collects the 50 observations over five days and plots the data in Figure 13.19. The mean is 1.62 and the standard deviation 1.00.

c. Calculate the upper and lower control limits, and draw them on the chart.

d. What does the control chart tell us?

e. Which of the following options should Jill accept and which should she reject in her search for improvement? For each option, give a reason for your decision.

   i. Look at why the orders on the last day have all been late.

   ii. Collect more data and break down orders by month, day of week, time of day, type of order, etc., to determine any patterns.

   iii. Purchase a new production control system to help control production start dates and the ordering of components.

   iv. Get senior management to send a memo to all production workers making them aware of the importance of keeping to planned production dates.

   v. Phone customers and offer them discounts for placing orders more than five days ahead of delivery date.

5. Zia Sims is in charge of dispatch, where she audits customer complaints. She has recently heard concerns about the printing quality of cardboard boxes supplied to customers, so she passed this information back to the printing department. Jo Arimu, the printing manager, decides to monitor the proportion of boxes with printing defects supplied to customers, with a view to improving quality. Jo sets up a team from the department to look into the matter.

Sample	Number of defects	Sample	Number of defects	Sample	Number of defects
1	4	9	5	17	2
2	8	10	3	18	5
3	5	11	5	19	3
4	6	12	6	20	4
5	4	13	7	21	2
6	3	14	4	22	0
7	0	15	4	23	4
8	4	16	5	24	5

Figure 13.20   Printing defects on cardboard boxes.

a. What should they do before they start collecting the data?
b. Later they collect data on the number of defective boxes in 24 samples of 100 boxes taken from customers' orders. The data are given in Figure 13.20. Calculate the upper and lower control limits, and plot them on a control chart.
c. What can you say about the process? What would you do next?
d. The printing team use the control chart to monitor the process, and the samples they take over the next few days have defective numbers of 5, 7, 4, and 12, respectively. Which of the following strategies would you adopt first in your search for improvement, and why?
   i. Stop the process immediately and see if you can identify any reason for the last high proportion of defective boxes. Look at the printing die, inks, operators, etc., to see if anything is different.
   ii. Have a word with the operator, tell him he should adjust the printing machine because the ink is coming out too fast and that he should be more alert in the future.
   iii. Look at all the defective boxes over the period. Classify them by colour of ink, size of print area, etc., and see if you can spot any patterns.
   iv. Order new ink from a different supplier. Try it out over a few days to see if it improves the printing quality.
   v. Since you have been considering buying a more modern printing press for a while now, bring the purchase forward and get it into operation as soon as possible.
6. The manager of a large regional hospital is concerned with the waiting times experienced by patients arriving at accident and emergency (A&E). Hospital policy is that no patient wait longer than 60 minutes before he or she sees a doctor. The data in Figure 13.21 give the length of time in minutes (to the nearest 5 minutes) that 50 patients wait after arriving at A&E before they see a doctor.

20	40	35	25	20	100	40	30	50	55
130	60	50	15	10	45	85	70	40	110
240	70	90	20	30	65	45	50	10	15
110	20	70	45	35	30	45	55	15	85
50	45	65	70	40	45	40	55	65	45

Figure 13.21 A&E waiting times.

    a. Draw a stem and leaf plot of the data, and show the specification limits on your plot.

    b. What percentage of data is beyond the specification limits?

7. Customers of a type of plastic wrapping require that it be at least 0.01 mm thick. A control chart alongside the process that makes this wrapping indicates that the process is in statistical control but that some of the wrapping is coming out of the process with thickness less than 0.01 mm. The machine has several controls: temperature of plastic, temperature of roller, pressure of injection, pressure of rollers, and mix of plastic granules (high or low recycled content). You have been asked to advise the improvement team on the strategies they might adopt to reduce variation.

    a. What sort of improvement strategies should they consider and in what order? Give some relevant examples.

    b. Outline how they might set up an experiment to identify the best settings for the process.

8. You are the production manager in a large company and wish to encourage line managers and machine operators to set up control charts for their processes. What are the three most important SPC concepts you would impart to these managers and operators so that their implementation of the charts stands the best chance of succeeding?

9. A team of line workers is looking at the thickness of corrugated board used to make cardboard boxes. Every hour they take three samples of board and measure the thickness of the board (in tenths of a millimetre), giving the results in Figure 13.22.

    a. Calculate the control limits, and set up the charts for monitoring the process.

    b. The specification limits for the width of board are 4.0–4.6 mm. What can you say about the capability of this process?

    c. You are responsible for reducing the variation in the thickness of the cardboard produced by the corrugation process. Give an example relating to this process of how you would investigate ways of reducing the variation in the thickness of the board.

    d. The charts have been in use for several months without the occurrence of any special causes. Under what circumstances could this be a problem?

10. Les Harrison, the mortgage manager of Northland State Bank, is concerned with the length of time that applicants for a mortgage wait to receive a decision from the bank. The bank's policy is that no one wait longer than six weeks

Hour	Sample 1	Sample 2	Sample 3	Mean	Range
1	43.45	42.07	43.61	43.04	1.54
2	40.37	43.32	42.28	41.99	2.95
3	41.66	47.49	46.03	45.06	5.83
4	48.08	41.98	43.40	44.49	6.10
5	41.35	40.63	39.03	40.34	2.32
6	38.03	47.28	43.90	43.07	9.25
7	46.32	38.43	44.58	43.11	7.89
8	40.82	45.76	40.31	42.30	5.45
9	42.78	43.54	40.02	42.11	3.52
10	48.13	43.55	42.40	44.69	5.73
11	39.77	43.26	42.81	41.95	3.49
12	41.66	45.47	42.76	43.30	3.81
13	41.84	42.34	40.93	41.70	1.41
14	47.21	42.26	48.65	46.04	6.39
15	43.11	41.95	40.81	41.96	2.30
16	47.26	44.07	41.86	44.40	5.40
17	42.31	35.96	43.69	40.65	7.73
18	42.32	46.00	43.66	43.99	3.68
19	44.63	42.22	46.10	44.32	3.88
20	41.84	46.87	47.53	45.41	5.69
21	43.96	40.05	47.32	43.78	7.27
22	45.39	44.89	46.54	45.61	1.65
23	43.89	43.86	45.29	44.35	1.43
24	43.62	39.86	43.05	42.18	3.76
Totals				1039.82	108.47

Figure 13.22 Thickness of corrugated board.

25	45	33	23	24	43	57	36	47	61
9	15	28	15	10	83	50	54	35	48
41	27	48	20	31	50	64	49	6	10
34	22	67	50	37	45	32	51	13	55
51	42	64	26	43	40	47	54	45	45

Figure 13.23 Waiting time for mortgage decision (days).

4	2	0	4	2	1	2	2	5	3
0	2	1	1	7	2	1	1	2	4
2	0	1	5	4	1	4	3	3	2
5	0	3	0	2	2	2	1	1	1
2	2	1	1	3	1	1	3	3	2

Figure 13.24   Accidents per week.

for the outcome of their mortgage application. The data in Figure 13.23 gives the number of days it took for 50 applicants to receive a decision.

a. Draw a stem and leaf plot of the data, and show the specification limits on your plot.

b. What percentage of data is beyond the specification limits?

c. The mean waiting time of the sample is 39 days, with a standard deviation of 17 days. What capability state is this process in?

11. Hill's Weaving Company is looking at two suppliers of yarn to be woven into men's suits. The elasticity of the yarn is an important attribute, because the yarn tends to snap if the elasticity is low. The elasticity of 40 samples of yarn has been measured from both suppliers. Supplier A's yarn has an elasticity with a mean of 28.3 and a standard deviation of 3.5, whereas supplier B's has a mean of 33.0 and a standard deviation of 5.5. If the elasticity is below 12, the yarn is unacceptable, because it will snap in use. In Exercise 4 in Chapter 4 you were asked which supplier you would choose and why. In the light of the material covered in the present chapter, would you change the supplier you previously recommended? Explain the reasons for your answer.

12. The road accidents in a specific area south of Auckland have been recorded over the last year (50 weeks), and the number of accidents per week is shown in Figure 13.24. The total number of accidents over the period is 107.

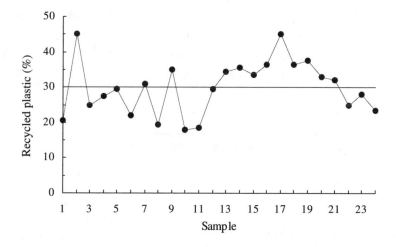

Figure 13.25   Percentage recycled plastic.

Day	Carton 1	Carton 2	Carton 3	Carton 4	Carton 5	Mean	Range
1	11.43	12.45	13.65	12.97	13.03	12.71	2.22
2	11.69	9.87	14.28	9.58	12.36	11.56	4.70
3	10.58	8.20	11.20	14.13	16.04	12.03	7.84
4	15.26	13.60	11.38	10.84	13.28	12.87	4.42
5	14.81	12.13	10.81	11.92	16.82	13.30	6.01
6	13.04	15.51	8.67	13.67	17.39	13.66	8.72
7	15.35	14.14	9.65	11.93	10.31	12.28	5.70
8	15.26	12.40	8.75	13.28	11.14	12.17	6.51
9	9.63	12.71	11.56	10.65	13.62	11.63	3.99
10	14.11	11.91	6.92	13.48	10.81	11.45	7.19
11	12.60	15.38	15.11	14.27	15.47	14.57	2.87
12	12.60	14.89	11.98	15.76	18.32	14.71	6.34
13	13.76	11.89	12.26	17.76	14.88	14.11	5.87
14	15.06	15.91	12.52	16.86	15.04	15.08	4.34
15	17.96	15.39	17.65	11.72	12.74	15.09	6.24
16	12.85	12.12	16.44	15.26	14.28	14.19	4.32
17	17.46	15.67	10.02	15.00	12.99	14.23	7.44
18	17.41	12.90	14.02	18.34	13.27	15.19	5.44
19	14.95	16.35	12.53	12.59	13.88	14.06	3.82
20	14.51	13.69	16.07	14.89	17.00	15.23	3.31
21	10.41	16.70	17.30	17.34	13.69	15.09	6.93
22	18.55	11.88	17.14	12.19	15.06	14.96	6.67
23	13.46	13.75	14.54	13.05	15.71	14.10	2.66
24	12.82	14.31	13.10	13.93	12.43	13.32	1.88
25	11.42	14.98	14.40	15.54	14.70	14.21	4.12
26	12.46	15.73	12.63	15.47	13.73	14.00	3.27
27	14.45	17.21	17.41	17.61	15.64	16.46	3.16
28	10.96	8.50	13.07	16.88	12.22	12.33	8.38
29	11.55	13.53	11.30	17.60	13.13	13.42	6.30
30	14.80	13.51	15.40	14.50	16.73	14.99	3.22
Totals						413.00	153.88

Figure 13.26   Fat content of yogurt cartons.

The highway authority is trying to reduce the accidents in this area to below three per week. Accidents per week over this specification limit are to be investigated, and traffic departments in the area will have to report on the causes to the head office in Wellington.

a. Is this a good strategy? What are some of the possible consequences of such a policy?

Hour	Sample 1	Sample 2	Sample 3	Sample 4	Mean	Range
1.00	20.74	20.28	23.43	19.55	21.00	3.88
2.00	20.87	21.33	20.96	18.57	20.43	2.76
3.00	21.45	20.04	22.50	18.89	20.72	3.61
4.00	28.24	24.11	18.70	22.08	23.28	9.54
5.00	17.61	23.64	19.52	16.15	19.23	7.49
6.00	17.65	20.75	19.21	15.99	18.40	4.76
7.00	20.73	19.65	18.60	20.57	19.89	2.13
8.00	17.87	20.50	18.79	22.40	19.89	4.53
9.00	21.50	24.73	20.77	23.69	22.67	3.96
10.00	22.69	19.38	21.89	16.68	20.16	6.01
11.00	19.58	21.30	17.78	17.98	19.16	3.52
12.00	17.92	22.91	22.20	18.86	20.47	4.99
13.00	22.01	18.63	22.77	23.08	21.62	4.45
14.00	19.87	19.55	23.28	17.72	20.11	5.56
15.00	19.91	20.26	21.46	21.98	20.90	2.07
16.00	17.83	22.59	19.67	23.15	20.81	5.32
17.00	21.88	22.35	19.94	19.14	20.83	3.21
18.00	18.92	23.81	21.67	20.39	21.20	4.89
19.00	20.82	19.22	21.19	16.84	19.52	4.35
20.00	20.42	19.36	16.34	17.48	18.40	4.08
Total					408.69	91.11

Figure 13.27   Depth of resin on tabletops.

      b. Which control chart is most appropriate for these accident data?
      c. Construct control limits for this set of data.
      d. If the limit 3 is treated as an upper specification limit, what capability state is this process in?
      e. What sort of actions would you recommend to try to reduce the accident toll in this area?

13. A plastics company is looking into the variability of the percentage content of recycled plastic used for moulding plastic buckets. Customers have been returning buckets that crack too easily when dropped. The company believes this is due to the recycled plastic content of the buckets, which should never be more than 40%. Figure 13.25 gives a run chart of the percentage of recycled plastic measured from 24 samples taken from the feed hopper at 10-minute intervals over one shift.
      a. Comment on the state of the process.
      b. What steps would you take in investigating this problem?

14. Brook Farm Ltd. is having trouble with the fat content of its fresh fruit yogurt. Customers have been complaining about fat globules in their cartons. To investigate this problem, Jim Donald, the production manager, decides to

collect samples from the yogurt process and measure the percentage fat content in order to set up control charts. The specification limits for the fat content are 15% ± 5%. Jim thinks that he will sample five cartons from the process every day for the next 30 days of production and measure the percentage fat content of each carton. The data Jim collects over the 30 days are given in Figure 13.26.

  a. Why do you think this is a suitable approach to establishing the control chart?

  b. Use the data to construct appropriate control charts.

  c. Is the process in statistical control? What do you think has happened to the process?

  d. What is the capability state of the process?

  e. How would you go about trying to reduce the variation in the fat content of the cartons of yogurt? Give some examples of the type of things you would do to reduce variation in this particular case.

15. The production manager of KC Kitchens Ltd. is looking at the final depth of an epoxy resin that is applied to a kitchen tabletop. Every hour she takes four samples and measures the depth (in hundredths of a millimetre). Figure 12.27 shows the measurements over 20 hours.

  a. Set up the control charts for this process.

  b. The specification limits for the depth of the resin are 16–24 (mm/100). What can you say about the capability of this process?

  c. What sort of strategies would you take to improve this process? List some ways in which you could apply these strategies to this process.

Postscript

Not all statistical topics of importance to managers and future managers can be covered in a book of this nature. Nor is it appropriate to do so. One of our aims has been to introduce you to methods that are or should be widely used in business. More importantly, our primary aim has been to get you thinking statistically about some of the issues faced in business. For some, more training may be needed. For instance, economists should further study regression (econometric methods) and forecasting. Those working in marketing need more on sampling methods and on techniques that help them understand the factors underlying consumer preferences. Production managers will benefit from a greater understanding of statistical process control and of the enormous benefits that can arise from using well designed experiments.

Decisions Based on Data

Throughout the book we have stressed the importance of basing decisions on data. You cannot understand variability unless you know the extent of the variability in the process you are studying. For this you need data. You cannot effectively improve a process or solve a problem unless you know the current state of the process or the scale of the problem. Again you need data. How can you understand the relationship between variables unless you look at the data? Useful forecasts of sales, unemployment, or commodity prices, for instance, cannot be made unless you know what has happened in the immediate past. Deciding whether a process is capable of meeting customer needs or whether to apply common cause or special cause strategies has to be based on an analysis of relevant data.

Too often such decisions are made on the basis of someone's hunch or gut feeling. There is nothing wrong with having some hunch about a process or a problem, based on some previous experience perhaps, but it needs to be supported and backed by data. Otherwise there may be no basis for the decision, and this can prove to be very costly if at some later time the hunch turns out to be misguided, as is often the case.

Thinking Statistically: The Key Principles

Very often managers find it difficult to know how best to use data or specify what data to collect. Often data are collected without a specific purpose, and much of it is never used. At other times data are used that have been collected for some other purpose and are not really suitable for the problem or process under study. The following general principles will help you decide what data to collect and how to collect, present, and analyse it.

Define Clearly What Is Being Studied

If you are trying to improve a process because too many errors are being made, variability is too great, or customers are complaining, then you should identify clearly what that process involves. Where does it begin and end? What are the stages involved in the process? Who or what is involved in the process? Simple top-down flowcharts are particularly useful in helping people understand a process or processes (see Section 2.5). If good forecasts of sales are required for production planning purposes, be clear about what it is you have to forecast. Is it the sales of a particular product line, of all products, or what? If you plan to survey your customers think carefully about what you want to find out. Planning what you are trying to achieve is a vital first step in any study or investigation.

Have Good Operational Definitions

Understand clearly and precisely what you want to observe and measure. This is not always easy, but it is essential if consistency in data is to be achieved. Decide how the data are going to be recorded. Use a well designed and tested *check sheet*. Make sure the equipment used to collect data is adequate for the job. One company installing double-glazing sent one of its new salespeople to measure the windows in a house. The person had not been told to measure the external, rather than the internal, dimensions of the window. When the new windows arrived, many of them did not fit. They had to be taken back to the factory and reworked. Essentially, it comes down to being adequately prepared to record or collect the required data by knowing precisely what you want to observe or measure and having the necessary equipment and facilities to do it.

Decide How to Collect the Data You Need

Usually it is not possible to collect information about every member of the population under study or of all the items produced by some process. A sample of members or items is needed. In Chapter 6 the wood experiment was used to demonstrate the potential dangers of taking a judgement sample. No matter how conscientious or careful he or she is, it is almost impossible for an investigator to use judgement to choose a sample that is free of bias. Increasing the sample size simply makes things worse, in that the bias is likely to increase.

Further, nothing can be said about the properties of the sample. With a random sample, however, we can determine these properties. For instance, we know that the distribution of the sample mean is approximately normal, is unbiased in the sense that the sample mean can be expected to equal the population mean in the long run and that the standard error of the sample mean decreases as sample size increases. Thus, we can attach a margin of error to our estimate using the methods described in Chapter 7. Different ways of taking a random sample were briefly mentioned in Chapter 6. Although nonrandom sampling is not recommended, it is frequently used in practice because, in many situations, random samples are difficult or costly to obtain. For example, market research data often derive from nonrandom samples, such as quota samples and self-selected samples. Data collected through a nonrandom sample are not necessarily worthless. It depends on the purpose of the study. Hotels, for instance, invariably leave a questionnaire in rooms asking for comments about the facilities and services provided. Information gained from such questionnaires can be useful in helping the hotel to better satisfy their customers' needs, even though the responses to such questionnaires do not necessarily constitute a random or representative sample of the hotel's customers.

Another important consideration in collecting data through a random sample is the question of nonresponse. A university carried out a survey of all its postgraduate students. Students were sent a questionnaire by e-mail and asked to fill it in and return it by e-mail. Only 28% of the students responded. What can we say about the 72% of students who did not respond? Were they the same as the group who did respond? Did they hold the same views? We do not know. The university may have been better advised to take a small random sample, stratified by factors such as gender and school of study, and follow up the initial e-mail questionnaire with further e-mails and personal interviews.

Another method of collecting data is through *experimentation*. Some of the principles involved were discussed in the context of the helicopter experiment in Chapter 13. Essentially an experiment involves some deliberate changes to the values, or levels, of factors that might have an important bearing on the outputs of the process. A careful choice of how these changes are made and applying them in a random order can give valuable insights into the probable effects of such changes. In an experiment the choice of which changes to make is under the control of the experimenter. In some situations, we are not free to make this choice. In an investigation of the relationship, if any, between smoking and lung cancer it would be unacceptable to choose which subjects smoked and which did not. All that can be done is to observe who smoked and whether they have lung cancer or not. This is an *observational study*, which can be useful for highlighting relationships and identifying possible causes. Only a properly designed and run experiment can demonstrate likely causation.

Collect Interval Data Whenever Possible

Greater information can be obtained from interval data than from attribute data. As explained in Chapter 5, the only property that attribute data possess is that of equivalence between items. It is not possible to say that one data value is so many units more than or less than another value. However, attribute data are usually quick and easy to collect. Avoid categorising interval data before analysing them. Knowing that a manufactured part is defective is not as informative as knowing the actual measurements of the part. Two defective parts can have very different characteristics, from barely unacceptable to totally unacceptable.

Think Carefully How to Present the Data

In the first week of each month a report on the costs of poor quality was presented to a general manager. Nothing was ever done about these costs because the data were presented in such an unintelligible way that the general manager failed to appreciate the importance of the figures. She was, after all, a busy person. It would have been better to bring attention to the main features of the data in a few well-presented tables and graphs; all the data could be placed at the end in an appendix, if need be. Advice on how to present data is given in Chapter 3, and two excellent books on this topic are listed in the bibliography. Reading these books will be well worthwhile.

Plot Your Data

The analysis of data should always begin by first plotting the data to look for any obvious patterns or relationships. With attribute data, use bar charts and Pareto diagrams. With interval data, a stem and leaf plot or histogram will tell you about the location, spread, and shape of the data. A box plot is useful for comparing sets of interval data. These plots will also draw attention to unusual data points (outliers), or points that are different from the remainder of the data. Such points may be the result of mistakes and errors, which can usually be rectified, or they could be important genuine points. In this latter situation they would normally be analysed separately from the remainder of the data. An example is given by the debt collection data discussed in Section 1.3 and elsewhere. Here, the evidence seems clear that the January figures are different from those of the other months of the year. If the data have been collected in some time sequence, then run charts should also be used, because patterns and trends over time, which are almost always important, are more easily detected. In studying relationships between variables, as in Chapters 8 and 9, scatterplots are essential. They indicate the type of regression model to fit and provide a means of determining the appropriateness of the chosen model. Keep in mind that different graphs show different aspects of the data. Sometimes a run chart shows an important pattern in the data; in other instances a stem and leaf plot, focusing on the shape of the data, will give a better understanding.

Keep It Simple

In collecting and plotting the data you may find that further analysis of the data is unnecessary. A Pareto diagram, for instance, will show where further work should and should not be concentrated. A check sheet, especially one with a visual element, may tell you where the problem lies without the need for any additional analysis. Some of the time, however, it will be necessary to carry out further analyses. It is important at this stage to keep in mind the purpose of your investigation. What you are trying to achieve in the analysis should have been identified at the outset. In a market research survey, are you really interested in discovering any difference in the preferences between female and male customers? If not, then an analysis of any difference becomes unnecessary. If you are, then what decisions will be made if a real difference is found?

Take Advice When Necessary

Numerous methods of analysis have been presented, for both attribute and interval data, and will be adequate for many of the situations that arise in business. In some cases more advanced methods will be required. Advice from a statistician or further training will then be necessary.

What We Have Not Done

In many ways this book is deceptively simple. It does not contain many formulae, and it uses algebra only where necessary to represent a difficult concept succinctly. Do not be misled by this. In many ways, this book is more advanced than those that only show you how to manipulate formulae. Doing routine calculations and blindly using formulae that you do not really understand might look clever, but at best it is rather pointless, and at worst potentially dangerous. You would not use a piece of complex machinery or drive a car without understanding what you are doing and why you are doing it. Likewise, you should not use powerful statistical tools without this understanding. It is important to know how to use data sensibly and to know what not to do, either because it is inappropriate or because you lack the necessary skills.

In Chapter 7 our emphasis was on estimating population parameters from sample data. In our opinion, most managers are or should be much more concerned with problems arising from the estimation of unknown population characteristics than in testing rather meaningless hypotheses. When trying to understand and explain variability we are, however, interested in examining whether differences we observe between factors in the process are meaningful, or significant, or whether such differences are due entirely to chance — in other words, in identifying factors that can explain this variability. In Chapter 5 we used the chi-square test to examine whether apparent differences between attribute factors were real or not, for example, whether females had a greater preference for a product than males. In regression analysis we try to explain the variation in the response variable by one or more explanatory variables. In Chapters 8 and 9 we used R^2 for this purpose. More advanced techniques for examining whether an explanatory variable or set of variables is contributing significantly can be found in any book on regression analysis. When the explanatory variables are attribute variables, we showed that dummy variables could be used in the regression analysis. However, separate methods exist for analysing data where all or almost all of these variables are attribute variables. These so-called *analysis of variance* methods are very powerful, but they are beyond the scope of this book. In the helicopter experiment, for example, we can use analysis of variance techniques to explain some of the variability in flying time in terms of the six factors in the experiment.

We do not think it necessary to explore the many aspects of formal probability theory or standard probability distributions that form the core of statistical theory. The one exception is the normal distribution. Due to its central role in statistical theory, it is important to fully understand the nature and purpose of this distribution. As to other distributions, however, we have made use of the key ideas from the t-, chi-square, binomial, and Poisson distributions, and this is far more important than being able to perform largely meaningless probability calculations. This is the domain of the statistician rather than the manager.

We have chosen not to include certain areas because we consider them unnecessary for the purpose of this book. However, many other areas exist where valuable insights into complex data sets could be obtained from more advanced statistical analysis but that are beyond the scope of this book. This is mainly in the fields of multivariate analysis and more sophisticated analyses of time series. The problems of understanding complex relationships between large numbers of variables and in being able to detect patterns and trends in data recorded over long periods of time are the two main areas where the advice of a statistician is strongly recommended.

Statistical Computing

Computers are very good at doing the technical work. You do not have to know the mathematical formula for the standard deviation in order to calculate it. It is easily done on the computer, using software such as Excel. What is more important is that you understand what it is and how to interpret and use it. Multiple regression models are easy to fit with appropriate computer software, even though the mathematics behind the technique are very complicated. Again it is sufficient to understand what multiple regression does, how to use it, the idea of fitting a model by least squares, and how to check the adequacy of the model and the assumptions made. We have chosen Excel in this book because it is simple to use, and our experience has shown that this (or similar spreadsheet packages) is what most managers use in business. For most of the material we have covered it is more than adequate. You can easily enter data into a spreadsheet, draw graphs and tables, and carry out many different analyses. Further, the outputs from Excel can be readily copied into a word processing package when writing a report or into a graphics presentation package.

Specialised statistical packages are also available that will do everything covered in this book and a lot more. One that is widely used for teaching at the level of this book is *Minitab*. Again data are entered into a spreadsheet, although calculations are carried out not in the spreadsheet, but in a separate window (or worksheet). Probably the most widely used statistics package for the analysis of survey data is *SPSS* (*Statistics Package for the Social Sciences*). Like Minitab, *SPSS* also stores data in a rectangular array of rows and columns, much like a spreadsheet, but without the dynamic updating features of a normal spreadsheet. *SPSS* is ideal for the analysis of the very large datasets that arise from market research and other surveys. It contains a wide range of techniques for analysing complex multivariate models and is available in an easy to use Windows version.

Appendix

Statistical Tables

A.1 Normal Distribution

The table below gives the probability (P) that an item is greater than a value z for a standard normal distribution.

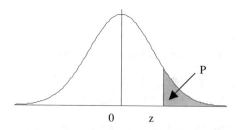

z	0.00	0.01	0.02	0.03	0.04	0.05	0.06	0.07	0.08	0.09
0.0	0.5000	0.4960	0.4920	0.4880	0.4840	0.4801	0.4761	0.4721	0.4681	0.4641
0.1	0.4602	0.4562	0.4522	0.4483	0.4443	0.4404	0.4364	0.4325	0.4286	0.4247
0.2	0.4207	0.4168	0.4129	0.4090	0.4052	0.4013	0.3974	0.3936	0.3897	0.3859
0.3	0.3821	0.3783	0.3745	0.3707	0.3669	0.3632	0.3594	0.3557	0.3520	0.3483
0.4	0.3446	0.3409	0.3372	0.3336	0.3300	0.3264	0.3228	0.3192	0.3156	0.3121
0.5	0.3085	0.3050	0.3015	0.2981	0.2946	0.2912	0.2877	0.2843	0.2810	0.2776
0.6	0.2743	0.2709	0.2676	0.2643	0.2611	0.2578	0.2546	0.2514	0.2483	0.2451
0.7	0.2420	0.2389	0.2358	0.2327	0.2296	0.2266	0.2236	0.2206	0.2177	0.2148
0.8	0.2119	0.2090	0.2061	0.2033	0.2005	0.1977	0.1949	0.1922	0.1894	0.1867
0.9	0.1841	0.1814	0.1788	0.1762	0.1736	0.1711	0.1685	0.1660	0.1635	0.1611
1.0	0.1587	0.1562	0.1539	0.1515	0.1492	0.1469	0.1446	0.1423	0.1401	0.1379
1.1	0.1357	0.1335	0.1314	0.1292	0.1271	0.1251	0.1230	0.1210	0.1190	0.1170
1.2	0.1151	0.1131	0.1112	0.1093	0.1075	0.1056	0.1038	0.1020	0.1003	0.0985
1.3	0.0968	0.0951	0.0934	0.0918	0.0901	0.0885	0.0869	0.0853	0.0838	0.0823
1.4	0.0808	0.0793	0.0778	0.0764	0.0749	0.0735	0.0721	0.0708	0.0694	0.0681
1.5	0.0668	0.0655	0.0643	0.0630	0.0618	0.0606	0.0594	0.0582	0.0571	0.0559
1.6	0.0548	0.0537	0.0526	0.0516	0.0505	0.0495	0.0485	0.0475	0.0465	0.0455
1.7	0.0446	0.0436	0.0427	0.0418	0.0409	0.0401	0.0392	0.0384	0.0375	0.0367
1.8	0.0359	0.0351	0.0344	0.0336	0.0329	0.0322	0.0314	0.0307	0.0301	0.0294
1.9	0.0287	0.0281	0.0274	0.0268	0.0262	0.0256	0.0250	0.0244	0.0239	0.0233
2.0	0.0228	0.0222	0.0217	0.0212	0.0207	0.0202	0.0197	0.0192	0.0188	0.0183
2.1	0.0179	0.0174	0.0170	0.0166	0.0162	0.0158	0.0154	0.0150	0.0146	0.0143
2.2	0.0139	0.0136	0.0132	0.0129	0.0125	0.0122	0.0119	0.0116	0.0113	0.0110
2.3	0.0107	0.0104	0.0102	0.0099	0.0096	0.0094	0.0091	0.0089	0.0087	0.0084
2.4	0.0082	0.0080	0.0078	0.0075	0.0073	0.0071	0.0069	0.0068	0.0066	0.0064
2.5	0.0062	0.0060	0.0059	0.0057	0.0055	0.0054	0.0052	0.0051	0.0049	0.0048
2.6	0.0047	0.0045	0.0044	0.0043	0.0041	0.0040	0.0039	0.0038	0.0037	0.0036
2.7	0.0035	0.0034	0.0033	0.0032	0.0031	0.0030	0.0029	0.0028	0.0027	0.0026
2.8	0.0026	0.0025	0.0024	0.0023	0.0023	0.0022	0.0021	0.0021	0.0020	0.0019
2.9	0.0019	0.0018	0.0018	0.0017	0.0016	0.0016	0.0015	0.0015	0.0014	0.0014
3.0	0.0013	0.0013	0.0013	0.0012	0.0012	0.0011	0.0011	0.0011	0.0010	0.0010

A.2 Percentage Points of the Normal Distribution

The table below gives the 90%, 95%, 98%, 99%, and 99.5% points of the standard normal distribution.

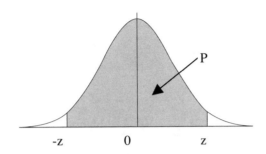

P	z
90%	1.645
95%	1.960
98%	2.326
99%	2.576
99.5%	2.807

A.3 Percentage Points of the *t*-Distribution

The table below gives the 90%, 95%, 98%, and 99% points of the *t*-distribution for different degrees of freedom (*df*).

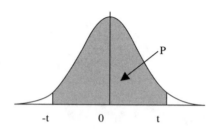

df	P = 90%	P = 95%	P = 98%	P = 99%
1	6.314	12.706	31.821	63.656
2	2.920	4.303	6.965	9.925
3	2.353	3.182	4.541	5.841
4	2.132	2.776	3.747	4.604
5	2.015	2.571	3.365	4.032
6	1.943	2.447	3.143	3.707
7	1.895	2.365	2.998	3.499
8	1.860	2.306	2.896	3.355
9	1.833	2.262	2.821	3.250
10	1.812	2.228	2.764	3.169
11	1.796	2.201	2.718	3.106
12	1.782	2.179	2.681	3.055
13	1.771	2.160	2.650	3.012
14	1.761	2.145	2.624	2.977
15	1.753	2.131	2.602	2.947
16	1.746	2.120	2.583	2.921
17	1.740	2.110	2.567	2.898
18	1.734	2.101	2.552	2.878
19	1.729	2.093	2.539	2.861
20	1.725	2.086	2.528	2.845
21	1.721	2.080	2.518	2.831
22	1.717	2.074	2.508	2.819
23	1.714	2.069	2.500	2.807
24	1.711	2.064	2.492	2.797
25	1.708	2.060	2.485	2.787
26	1.706	2.056	2.479	2.779
27	1.703	2.052	2.473	2.771
28	1.701	2.048	2.467	2.763
29	1.699	2.045	2.462	2.756
30	1.697	2.042	2.457	2.750
31	1.696	2.040	2.453	2.744

df	P = 90%	P = 95%	P = 98%	P = 99%
32	1.694	2.037	2.449	2.738
33	1.692	2.035	2.445	2.733
34	1.691	2.032	2.441	2.728
35	1.690	2.030	2.438	2.724
36	1.688	2.028	2.434	2.719
37	1.687	2.026	2.431	2.715
38	1.686	2.024	2.429	2.712
39	1.685	2.023	2.426	2.708
40	1.684	2.021	2.423	2.704
41	1.683	2.020	2.421	2.701
42	1.682	2.018	2.418	2.698
43	1.681	2.017	2.416	2.695
44	1.680	2.015	2.414	2.692
45	1.679	2.014	2.412	2.690
46	1.679	2.013	2.410	2.687
47	1.678	2.012	2.408	2.685
48	1.677	2.011	2.407	2.682
49	1.677	2.010	2.405	2.680
50	1.676	2.009	2.403	2.678
55	1.673	2.004	2.396	2.668
60	1.671	2.000	2.390	2.660
65	1.669	1.997	2.385	2.654
70	1.667	1.994	2.381	2.648
75	1.665	1.992	2.377	2.643
80	1.664	1.990	2.374	2.639
85	1.663	1.988	2.371	2.635
90	1.662	1.987	2.368	2.632
95	1.661	1.985	2.366	2.629
100	1.660	1.984	2.364	2.626
120	1.658	1.980	2.358	2.617
∞	1.645	1.960	2.326	2.576

A.4 Percentage Points of the χ^2-Distribution

The table below gives the 10%, 5%, 2%, and 1% points of the χ^2-distribution for different degrees of freedom (*df*).

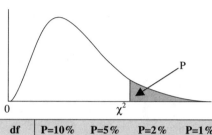

df	P=10%	P=5%	P=2%	P=1%
1	2.71	3.84	5.41	6.63
2	4.61	5.99	7.82	9.21
3	6.25	7.81	9.84	11.34
4	7.78	9.49	11.67	13.28
5	9.24	11.07	13.39	15.09
6	10.64	12.59	15.03	16.81
7	12.02	14.07	16.62	18.48
8	13.36	15.51	18.17	20.09
9	14.68	16.92	19.68	21.67
10	15.99	18.31	21.16	23.21
11	17.28	19.68	22.62	24.73
12	18.55	21.03	24.05	26.22
13	19.81	22.36	25.47	27.69
14	21.06	23.68	26.87	29.14
15	22.31	25.00	28.26	30.58
16	23.54	26.30	29.63	32.00
17	24.77	27.59	31.00	33.41
18	25.99	28.87	32.35	34.81
19	27.20	30.14	33.69	36.19
20	28.41	31.41	35.02	37.57
21	29.62	32.67	36.34	38.93
22	30.81	33.92	37.66	40.29
23	32.01	35.17	38.97	41.64
24	33.20	36.42	40.27	42.98
25	34.38	37.65	41.57	44.31
26	35.56	38.89	42.86	45.64
27	36.74	40.11	44.14	46.96
28	37.92	41.34	45.42	48.28
29	39.09	42.56	46.69	49.59
30	40.26	43.77	47.96	50.89
40	51.81	55.76	60.44	63.69
50	63.17	67.50	72.61	76.15
60	74.40	79.08	84.58	88.38
70	85.53	90.53	96.39	100.43
80	96.58	101.88	108.07	112.33
90	107.57	113.15	119.65	124.12
100	118.50	124.34	131.14	135.81

A.5 Random Numbers

70	71	42	34	1	74	88	50	53	94	22	16
52	5	30	51	18	11	4	67	22	80	13	32
69	2	3	77	29	67	40	41	100	13	92	19
98	78	82	48	43	52	19	75	98	75	88	23
73	3	53	81	72	95	65	71	1	69	2	21
86	19	20	95	98	53	32	87	79	1	35	12
47	29	13	71	23	41	17	88	9	59	52	61
76	16	8	67	66	82	26	34	33	91	88	31
72	9	60	10	90	94	35	80	26	70	69	98
13	4	4	40	81	89	34	69	39	56	36	47
5	57	88	29	89	28	66	44	94	48	38	3
43	97	3	7	87	11	14	20	71	39	62	31
59	52	29	4	92	14	85	66	50	44	57	44
9	40	24	18	24	20	82	58	66	48	76	23
40	66	45	90	2	36	16	96	52	58	4	65
92	99	65	71	67	43	21	51	96	59	44	6
86	30	89	97	15	55	24	70	43	56	43	87
46	48	40	74	9	70	39	48	85	58	85	41
79	85	9	38	96	11	25	89	43	75	6	27
52	57	35	28	71	63	68	67	57	58	60	48
4	69	6	36	89	8	22	58	10	80	83	61
90	83	78	47	26	52	2	9	36	20	42	87
24	76	52	91	92	60	35	9	84	22	46	33
22	56	83	19	69	40	91	60	74	31	15	91
12	17	30	86	49	70	75	94	59	89	73	85
21	75	2	45	32	69	49	65	68	71	94	62
47	84	67	37	37	41	64	55	64	33	47	74
74	56	36	86	29	76	3	80	39	7	28	45
85	54	82	33	42	32	65	82	19	44	18	18
93	7	68	14	54	51	47	44	11	47	13	2
3	99	36	18	25	99	42	23	84	98	85	84

A.6 Constants for Control Charts

Subgroup size	A_2	D_3	D_4
2	1.880	0	3.267
3	1.023	0	2.574
4	0.729	0	2.282
5	0.577	0	2.114
6	0.483	0	2.004
7	0.419	0.076	1.924
8	0.373	0.136	1.864
9	0.337	0.184	1.816
10	0.308	0.223	1.777

Bibliography and Further Reading

Three books we have referred to in the book and that we strongly recommend as giving the wider picture are:

B.L. Joiner. *Fourth Generation Management*. McGraw-Hill: New York, 1994.

P.R. Scholtes et al. *The Team Handbook*. Oriel Consulting: Madison, WI, 1988.

P.R. Scholtes. *The Leader's Handbook*. McGraw-Hill: New York, 1998.

Many books have been written about the management methods of W Edwards Deming and are well worth studying. As a start, we suggest:

W.E. Deming. *Out of the Crisis*. MIT Center for Advanced Engineering Study: Cambridge, MA, 1986.

M. Walton. *The Deming Management Method*. Putnam: New York, 1986.

Short introductory statistics texts focused on quality improvement that will supplement the material in Chapters 2, 3, and 11–13 in particular are:

K. Ishikawa. *Guide to Quality Control*. Asian Productivity Organisation: Tokyo, 1982.

H. Kume. *Statistical Methods for Quality Improvement*. AOTS Chosakai: Tokyo, 1985.

Two books recommended in Chapter 3 for their advice and wisdom on how to effectively present data are:

A.S.C. Ehrenberg. *A Primer in Data Reduction*. Wiley: New York, 1982.

E.R. Tufte. *The Visual Display of Quantitative Information*. Graphics Press: Cheshire, UK, 1983.

Many books have been written on the analysis of attribute data, usually under the guise of categorical data analysis or contingency table analysis. A good text to start with is:

A. Agresti. *An Introduction to Categorical Data Analysis*. Wiley: New York, 1996.

For those interested particularly in market research and requiring a fuller discussion of the principles and methods of survey sampling, we recommend:

V. Barnett. *Sample Survey Principles and Methods*. Edward Arnold: London, 1991.

The classic text on regression analysis, which was first published in 1966 and is now in its third edition, is:

N.R. Draper and H. Smith. *Applied Regression Analysis*. Wiley: New York, 1998.

There are a number of good books on forecasting aimed at business practitioners but often running to many hundreds of pages. For further study we recommend:
P.H. Franses. *Time Series Models for Business and Economic Forecasting*. Cambridge University Press: Cambridge, UK, 1988.
J.T. Mentzer and C.C. Bienstock. *Sales Forecasting Management: Understanding the Techniques, Systems, and Management of the Sales Forecasting Process*. Sage Publications: Newberry Park, CA, 1998.
S. Makridakis, S.C. Wheelwright, and R. J. Hyndman. *Forecasting. Methods and Applications*. Wiley: New York, 1998.

A more statistical introduction to time series is given in:
C. Chatfield. *The Analysis of Time Series: An Introduction*. Chapman and Hall: London, 2003.

A short but excellent introduction to statistical process control and its role in helping us understand variation is:
D.J. Wheeler. *Understanding Variation: The Key to Managing Chaos*. SPC Press: Knoxville, 1993.

For a more detailed treatment of statistical process control we recommend:
D.J. Wheeler and D.S. Chambers. *Understanding Statistical Process Control*. SPC Press: Knoxville. 1992.

The helicopter exercise in Section 13.7 gives a brief introduction to industrial experimentation. The following book provides an excellent introduction to the many issues and concepts involved in experimentation, as well as providing details on how to design and analyse experiments:

R.D. Moen, T.W. Nolan, and L.P. Provost. *Quality Improvement through Planned Experimentation*. McGraw-Hill: New York. 1999.

At a more advanced level, we recommend the following classic text:
G.E.P. Box, W.G. Hunter, and J.S. Hunter. *Statistics for Experimenters*. Wiley: New York, 1978.

Information on the statistical software packages *Minitab* and SPSS can be found on their Internet sites:

For *Minitab*   http://www.minitab.com/
For *SPSS*   http://www.spss.com/

Index